To the Memory of Günter Wenzel

Lecture Notes in Economics and Mathematical Systems

Managing Editors: M. Beckmann and W. Krelle

348

Horand Störmer

Binary Functions and their Applications

Springer-Verlag
Berlin Heidelberg New York London
Paris Tokyo Hong Kong Barcelona

Author

Prof. Dr. Horand Störmer
Lehrstuhl für Mathematik V
Fakultät für Mathematik und Informatik
Universität Mannheim, Seminargebäude A 5
D-6800 Mannheim 1, FRG

ISBN-13:978-3-540-52812-8 e-ISBN-13:978-3-642-61519-1
DOI:10.1007/978-3-642-61519-1

2142/3140-543210 – Printed on acid-free paper

Acknowledgements

The author is very grateful to Dipl.-Math. Birgit Schillinger and Dipl.-Math. Gerd Waldschaks for their careful and critical reading of these notes and many helpful suggestions and corrections.

He is very indebted to Gerda Jones for her skilful and talented work in producing this text from a handwritten manuscript.

He thanks Professor M. Beckmann for his willingness to publish these notes in this series, and the publishers for their agreeable cooperation.

Mannheim, April 1990 Horand Störmer

CONTENTS

1. Introduction 1

2. Binary Functions and their Representations by Implicants 3

2.1 Cube Indicators 4
2.2 Implicants 7
2.3 Prime Implicants 12
2.4 Representations by Implicants (Prime Implicants) 14

3. Representations of Binary Functions by Implicates 23

3.1 Anticube Indicators 23
3.2 Implicates 25
3.3 Prime Implicates 29
3.4 Representations by Implicates (Prime Implicates) 30

4. Reduction Methods 36

4.1 Notation 36
4.2 Rule 36
4.3 Rule 37
4.4 Rule 37
4.5 Rule 38
4.6 Rule 38
4.7 Rule (Corollary) 39
4.8 Rule 39
4.9 Rule 40

5. Discrete Functions 44

5.1 Representations by Binary Functions 44
5.2 Monotone Functions 47
5.3 Semimonotone Functions 52
5.4 Implicants (Prime Implicants) of Discrete Functions 56
5.5 Representations by Implicants (Prime Implicants) 61
5.6 Implicates (Prime Implicates) of Discrete Functions 68
5.7 Representations by Implicates (Prime Implicates) 74

6. Applications 80

6.1 Reliability Structure of Technical Systems 80
6.2 Classification (Valuation) of Objects 81

7. A Class of Finite Boolean Algebras 84

7.1 Boolean Algebras 85
7.2 Boolean Algebras Generated by Finite Partitions 88
7.3 Representations of Boolean Elements by Implicants 94
7.4 Representations of Boolean Elements by Implicates 99
7.5 Probability 102

8. Applications 110

8.1 Set Algebras (Event Algebras) 110
8.2 Indicator Algebras 117
8.3 Partitions in Propositional Logic 123
8.4 Algebras of Classes of Propositions 126
8.5 Truth Function Algebras 133
8.6 Some Related Models 136
8.7 Calculation of Elements of $\mathcal{B}^*$ 139

Concluding Remark 143

References 144

List of Symbols 145

Subject Index 149

Chapter 1

Introduction

The usefulness of Boolean functions

$$f : \{0,1\}^n \to \{0,1\}$$

within a lot of application fields is well–known and undisputed. Many books and papers are engaged with their several representations and their practical use in such important branches as computer science, probability theory or mathematical logic.

Boolean functions are special functions out of the wide class of discrete functions

$$f : \bigtimes_1^n M_i \to \mathbb{R}$$

where $M_1, ..., M_n$ are any finite subsets of $\mathbb{R}$.

In the following we consider binary functions

$$f : \bigtimes_1^n M_i \to \{0,1\},$$

i.e. special discrete functions containing the class of Boolean functions.

Binary functions are of particular interest: On the one hand they are easy to treat with respect to their representations by so–called implicants respectively implicates. On the other hand they are important for the representation of general discrete functions. Moreover implicants respectively implicates of discrete functions may be constructed with the aid of implicants respectively implicates of related binary functions. These results immediately lead to applications to some problems of reliability theory and classification theory.

Later we will show that the set of all binary functions defined on a fixed support $\bigtimes_1^n M_i$ may be interpreted as a special case of a class of Boolean algebras generated by n partitions of a unit element Ω. This allows to translate the results about binary functions to further models in probability theory and propositional logic.

In Chapter 2 we show first how binary functions may be represented by maxima of implicants of them. Implicants are defined as indicators of subsets $\bigtimes P_i$ of $\bigtimes M_i$. Among other things we show that every binary function has minimal representations by prime implicants, i.e. representations with a minimal number of implicants.

The representation of binary functions by minima (products) of implicates is treated in Chapter 3. Implicates are indicators of complements of subsets $\times P_i$ of $\times M_i$. Every binary function also has minimal representations by prime implicates, i.e. presentations with a minimal number of "minimal" implicates.

To construct the minimal representations of binary functions by implicants or implicates in Chapter 2 and Chapter 3 we need some reduction methods given in Chapter 4.

In Chapter 5 we first show that every discrete function may be represented in an evident way by suitable binary functions. Such representations are of particular interest in case of monotone discrete functions (used e.g. to describe the reliability structure of technical systems). Then we show how the problem to find the implicants respectively implicates of discrete functions may be reduced to the construction of implicants respectively implicates of binary functions.

Two substantial applications of binary and discrete functions with their representations by implicants and implicates are given in Chapter 6. First we consider reliability structures of technical systems with more then two states. Then we use binary functions respectively discrete functions to classify objects which are described by the grades of n certain attributes.

The structure of binary functions suggests to consider – in Chapter 7 – more generally a class of finite Boolean algebras generated by n partitions of the unit element Ω. Each element of such a finitely generated Boolean algebra has representations by implicants and implicates deduced immediately from the implicants and implicates of a corresponding binary function. Moreover we may define a probability measure on the generated Boolean Algebra.

Finally in Chapter 8 the results of Chapter 7 will be used to give representations of the sets (events) of a finitely generated set algebra by implicants and implicates. Furthermore we may give such representations for indicator functions, classes of logical propositions and truth functions.

Chapter 2

Binary Functions and their Representations by Implicants

For fixed $n \in \mathbb{N}$ and $i = 1, ..., n$ let

$$M_i := \{a_{io}, ..., a_{ik_i}\}, k_i \in \mathbb{N}, a_{io} < \cdots < a_{ik_i}$$

be any subsets of $\mathbb{R}$ and $M := \bigtimes_1^n M_i$. We consider functions

$$f : M \to \{0, 1\}$$

and call them *binary functions*. Thus a binary function is a function of n finite-valued variables and takes only the values 0 and 1.

Let $\Gamma := f^{-1}(\{1\}) := \{x \in M : f(x) = 1\} \subseteq M$. Then we may write f as the *indicator* (indicator function) of Γ by

$$f(x) = 1_\Gamma(x) := \begin{cases} 1 & \text{for } x \in \Gamma \\ 0 & \text{for } x \in \overline{\Gamma} := M \setminus \Gamma. \end{cases}$$

Conversely each indicator on M obviously is a binary function. Therefore in the following we always write binary functions as 1_Γ. To construct representations of 1_Γ by implicants first we need so-called cube indicators and their properties.

Preliminary Remark

Our definition of a binary function on any arbitrary cartesian product M is the most general one. We may simplify it by choosing $M_i := \{0, ..., k_i\}$ and so $M = K := \bigtimes_{i=1}^n \{0, ..., k_i\}$. Then we obtain a simpler notation, but this model may be not directly applicable in some cases.

On the other hand we may obtain the general results from the special results with $M = K$ in the following way.

1. We denote the elements of K by $j := (j_1, ..., j_n)$ with $j_1 \in \{0, ..., k_1\}, ...,$ $j_n \in \{0, ..., k_n\}$. The subsets of K are denoted by Γ.

For the elements of M we write $a_j := (a_{1j_1}, ..., a_{nj_n})$ with $j = (j_1, ..., j_n) \in K$ as before. Now to each $\Gamma \subseteq K$ we define the coordinated set $G_\Gamma \subseteq M$ by

$$G_\Gamma := \{a_j : j \in \Gamma\}.$$

Obviously, for $\mathcal{P}(M)$, the set of all subsets of M holds

$$\mathcal{P}(M) = \{G_\Gamma : \Gamma \in K\}.$$

The system $\mathcal{P}(M)$ is an algebra in M. From the definition of G_Γ it follows $(\Gamma, \Gamma' \subseteq K)$ $G_\Gamma \cup G_{\Gamma'} = G_{\Gamma \cup \Gamma'}, G_\Gamma \cap G_{\Gamma'} = G_{\Gamma \cap \Gamma'}, \overline{G_\Gamma} = G_{\overline{\Gamma}}$, further $G_\Gamma \subseteq G_{\Gamma'}$ if and only if $\Gamma \subseteq \Gamma'$, especially $G_\Gamma = G_{\Gamma'}$ if and only if $\Gamma = \Gamma'$, $G_\Gamma \subset G_{\Gamma'}$ if and only if $\Gamma \subset \Gamma'$.

2. Now consider to each $\Gamma \subseteq K$ the indicators (binary functions)

$$1_\Gamma : K \to \{0,1\} \text{ with } 1_\Gamma(x) = \begin{cases} 1 & \text{for } x \in \Gamma \\ 0 & \text{for } x \in \overline{\Gamma} \end{cases}$$

and

$$1_{G_\Gamma} : M \to \{0,1\} \text{ with } 1_{G_\Gamma}(x) = \begin{cases} 1 & \text{for } x \in G_\Gamma \\ 0 & \text{for } x \in \overline{G_\Gamma}. \end{cases}$$

Then it follows from the preceding that any relation

$$1_\Gamma = A(1_{\Gamma_1}, ..., 1_{\Gamma_r})$$

between the indicators $1_\Gamma, 1_{\Gamma_1}, ..., 1_{\Gamma_r}$ holds if and only if

$$1_{G_\Gamma} = A(1_{G_{\Gamma_1}}, ..., 1_{G_{\Gamma_r}})$$

holds.

For example the relation $1_\Gamma = \max(1_{\Gamma_1}, 1_{\Gamma_n})$ holds if and only if $1_{G_\Gamma} = \max(1_{G_{\Gamma_1}}, 1_{G_{\Gamma_2}})$ holds.

This means: we may obtain all results of Chapter 2 and 3 concerning binary function on M and their several representations (by implicants and implicates) from the results in case $M = K$. We only have to replace all sets $\Gamma \subset K$ by $G_\Gamma \subset M$ respectively all indicators 1_Γ with $\Gamma \subseteq K$ by the corresponding indicators 1_{G_Γ}.

In the following however we consider more generally each Γ as subset of M and so 1_Γ as binary function defined on M.

As for the particular meaning of the case $M = K$ see Chapter 7 and Chapter 8 concerning finitely generated Boolean algebras and their applications.

2.1 Cube Indicators

2.1.1 Definition

If $\emptyset \neq \bigtimes_1^n P_i \subset M$ we call $\bigtimes P_i := \bigtimes_1^n P_i$ a *cube* and $C(P) := C(P_1, ..., P_n) : M \to \{0,1\}$, defined by

$$C(P_1, ..., P_n) := 1_{\bigtimes P_i}$$

a *cube indicator.*

If $|P_i| = 1$ for $i = 1, ..., n$, i.e. $\times P_i = \{a\}, a \in M$, then we write $C(a)$ instead of $C(P)$ and call it a *minterm.*

If $\emptyset \neq K \subset N_n := \{1, ..., n\}$ and $|P_i| = 1$ for $i \in K$, $|P_i| > 1$ for $i \in \overline{K} := N_n \setminus K$, then we write $C(a^K, P^{\overline{K}})$ instead of $C(P)$ where $a^K \in \underset{K}{\times} M_i$. (Clearly if $K = \{j_1, ..., j_k\}$, $j_1 < \cdots < j_k$ then $a^K \in \underset{K}{\times} M_i$ means $a^K = (a_{j1}, ..., a_{jk})$ with $a_{j1} \in M_{j1}, ..., a_{jk} \in M_{jk}$.) Further we use $C(a^\phi, P^{N_n}) := C(P)$ (if $|P_i| > 1$ for $i \in N_n$) and $C(a^{N_n}, P^\phi) := C(a)$.

By its definition a cube indicator $C(P)$ takes the value 1 if and only if always $x_i \in P_i$ for $i = 1, ..., n$. Let $P_1, ..., P_n$ be given by $P_i = \{\alpha_{i1}, ..., \alpha_{im_i}\} \subseteq M_i$ for $i = 1, ..., n$. Then we may illustrate $C(P)$ by the following series circuit of n parallel circuits:

(2.1.2) C(P):

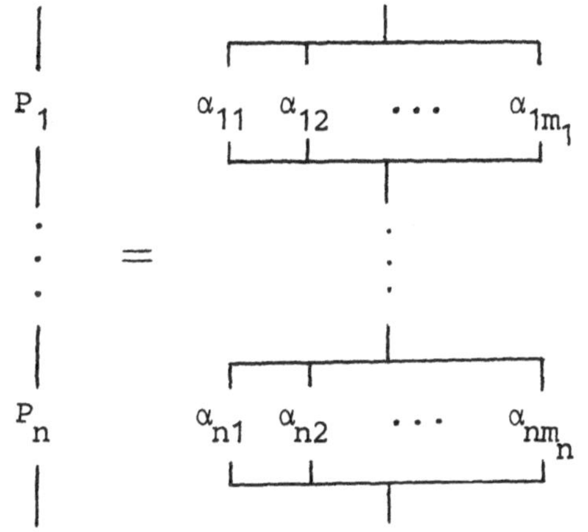

($C(P) = 1$ if and only if $x_1 \in P_1$ and $x_2 \in P_2$ ... and $x_n \in P_n$).

Now we consider relations between cube indicators. We may state the following result.

2.1.3 Lemma

(a) If $L \subset K$ then

$$C(a^L, P^{\overline{L}}) = \sum_{a^{K\setminus L} \in \bigtimes_{K\setminus L} P_i} C(a^K, P^{\overline{K}}) = \max_{a^{K\setminus L} \in \bigtimes_{K\setminus L} P_i} C(a^K, P^{\overline{K}}).$$

Clearly the sum resp. the maximum on the right hand side has to be taken over all $C(a^K, P^{\overline{K}})$ with fixed a^L (i.e. fixed a_i for $i \in L$), fixed $\bigtimes_{\overline{K}} P_i$ (i.e. fixed P_i for $i \in \overline{K}$) and all $a^{K\setminus L} \in \bigtimes_{K\setminus L} P_i$.

(b) The following are equivalent:

(α) $C(P) \leq C(Q)$.

(β) $\bigtimes P_i \subseteq \bigtimes Q_i$.

(c) The following are equivalent:

(α) $C(a^K, P^{\overline{K}}) \leq C(b^L, Q^{\overline{L}})$.

(β) $L \subseteq K, a^L = b^L$ (if $L \neq \phi$),
$a^{K\setminus L} \in \bigtimes_{K\setminus L} Q_i$, (if $L \subset K$), $\bigtimes_{\overline{K}} P_i \subseteq \bigtimes_{\overline{K}} Q_i$ (if $K \neq N_n$).

Proof

(a) For each cube $\bigtimes P_i$, we define the cubes $P_i^* \subseteq M$ for $i = 1, ..., n$ by

$$(2.1.4) \qquad P_i^* := \{x \in M : x_i \in P_i\} = \bigtimes_1^n R_j \text{ with } R_j := \begin{cases} P_i & \text{for } j = i \\ M_i & \text{for } j \neq i. \end{cases}$$

Then

$$\bigtimes P_i = \bigcap P_i^* \text{ and } C(P) = 1_{\bigcap P_i^*}.$$

Now the proposed equation is equivalent to the set relation

$$\bigcap_L \{a_i\}^* \cap \bigcap_{\overline{L}} P_i^* = \dot{\bigcup}_{a^{K\setminus L} \in \bigtimes_{K\setminus L} P_i} \left[\bigcap_{K\setminus L} \{a_i\}^* \cap \bigcap_L \{a_i\}^* \cap \bigcap_{\overline{K}} P_i^*\right]$$

(union of mutually disjoint sets, $\bigcap_\phi \cdots := M$) which is evident.

(b) This is obvious.

(c) This is a specification of (b). □

2.2 Implicants

Let us consider any binary function $1_\Gamma : M \to \{0,1\}$ (with $\phi \neq \Gamma \subset M$). First from

$$\Gamma = \{a \in M : a \in \Gamma\} = \bigcup_{a\in\Gamma}\{a\} \text{ and } 1_{\{a\}} = C(a)$$

we obtain the unique representation

$$1_\Gamma = \sum_{a\in\Gamma} C(a) = \max_{a\in\Gamma} C(a) \tag{2.2.1}$$

of 1_Γ by minterms $C(a)$ where $a \in \Gamma$ is equivalent to $C(a) \leq 1_\Gamma$.

Now we define implicants of 1_Γ as special cube indicators.

2.2.2 Definition

A cube indicator $C(P)$ is called an *implicant* of 1_Γ if $C(P) \leq 1_\Gamma$, equivalent to $\times P_i \subseteq \Gamma$. The implicants $C(a)$ (minterms) of 1_Γ according to (2.2.1) are called the *minimplicants* of 1_Γ.

Comment

Obviously a minterm $C(a)$ is a minimplicant of 1_Γ if and only if $a \in \Gamma$, and 1_Γ equals the sum (the maximum) of all its minimplicants. Further a cube indicator $C(P)$ is an implicant of 1_Γ if and only if $C(P)(x) = 1$ (i.e. $x_1 \in P_1, ..., x_n \in P_n$) always implies $1_\Gamma(x) = 1$.

Our definition of implicants generalizes the appropriate definition in the Boolean case corresponding to $M = \{0,1\}^n$. Then a cube always is of the form $\times P_i$ with $P_i \in \{\{0\},\{1\},\{0,1\}\}$ and for the cube indicator $1_{\times P_i}$ it holds

$$1_{\times P_i}(x) = \prod_1^n 1_{P_i^*}(x_i)$$

with

$$1_{P_i^*(x_i)} = \begin{cases} 1 - x_i & \text{if } P_i = \{0\} \\ x_i & \text{if } P_i = \{1\} \\ 1 & \text{if } P_i = \{0,1\} \end{cases} .$$

Each cube indicator thus may be written uniquely as

$$1_{\times P_i}(x) = \prod_{j \in N} x_j \prod_{j \in N'} (1 - x_j)$$

with $N, N' \subseteq N_n$, $N \cap N' = \emptyset$. Furthermore a cube evidently is always a convex cube, and so an implicant of a Boolean function is always the indicator of a convex cube.

In the general case we do not require $\times P_i$ to be convex, i.e. to be an interval. However it is possible to restrict the definition of implicants to interval indicators. Then we obtain some modifications in the following theory. On the other hand all results concerning interval indicators may be obtained by the methods of the general theory.

Now let us determine the set of all implicants of 1_Γ.

From Lemma 2.1.3 we obtain the following theorem concerning the set of all implicants of 1_Γ.

Denote by $C(\Gamma)$ the set of all implicants of 1_Γ. For $k = 0, ..., n$ define $C_k(\Gamma)$

$$C_k(\Gamma) := \{C(a^K, P^{\overline{K}}) \in C(\Gamma) : |K| = k\},$$

so that

$$C(\Gamma) = \bigcup_{o}^{n} C_k(\Gamma).$$

2.2.3 Theorem

Let $L \subset N_n$ with $|L| = k$. Then (for any cube indicator $C(a^L, P^{\overline{L}})$ the following are equivalent:

(a) $C(a^L, P^{\overline{L}}) \in C_k(\Gamma)$.

(b) For some K with $L \subset K \subseteq N_n, |K| = k'$ it holds:
$C(a^K, P^{\overline{K}}) \in C_{k'}(\Gamma)$ for each $a^{K \setminus L} \in \times_{K \setminus L} P_i$.

(c) For each K with $L \subset K \subseteq N_n, |K| = k'$ it holds:
$C(a^K, P^{\overline{K}}) \in C_{k'}(\Gamma)$ for each $a^{K \setminus L} \in \times_{K \setminus L} P_i$.

Proof

Lemma 2.1.3 (a) yields

$$C(a^L, P^{\overline{L}}) \leq 1_\Gamma \text{ (i. e. } C(a^L, P^{\overline{L}}) \in \mathcal{C}_k(\Gamma)) \Leftrightarrow$$
$$C(a^K, P^{\overline{K}}) \leq 1_\Gamma \text{ (i. e. } C(a^K, P^{\overline{K}}) \in \mathcal{C}_{k'}(\Gamma)) \text{ for each } a^{K\setminus L} \in \bigtimes_{K\setminus L} P_i. \quad \square$$

From Theorem 2.2.3 we deduce the following result describing the connection between $\mathcal{C}_k(\Gamma)$ and $\mathcal{C}_{k+1}(\Gamma)$.

2.2.4 Corollary

Let $L \subset N_n$ with $|L| = k$. Then for any cube indicator $C(a^L, P^{\overline{L}})$ the following are equivalent:

(a) $C(a^L, P^{\overline{L}}) \in \mathcal{C}_k(\Gamma)$.

(b) For some $j \in \overline{L}$ holds:
$C(a^{L\cup\{j\}}, P^{\overline{L}\setminus\{j\}}) \in \mathcal{C}_{k+1}(\Gamma)$ for each $a_j \in P_j$.

(c) For each $j \in \overline{L}$ holds:
$C(a^{L\cup\{j\}}, P^{\overline{L}\setminus\{j\}}) \in \mathcal{C}_{k+1}(\Gamma)$ for each $a_j \in P_j$.

Proof

This follows from Theorem 2.2.3 with $K = L \cup \{j\}$. $\square$

Corollary 2.2.4 in principle yields a method to gain stepwise $\mathcal{C}_{n-1}(\Gamma)$, $\mathcal{C}_{n-2}(\Gamma), \ldots, \mathcal{C}_o(\Gamma)$ beginning at the set $\mathcal{C}_n(\Gamma)$ of all minimplicants $C(a)$ of 1_Γ. Evidently if any $\mathcal{C}_k(\Gamma)$ is empty then also $\mathcal{C}_{k-1}(\Gamma)$. We note that according to Corollary 2.2.4 any $C(a^L, P^{\overline{L}})$ belongs to $\mathcal{C}_k(\Gamma)$ if and only if each $C(a^{L\cup\{j\}}, P^{\overline{L}\setminus\{j\}})$ with $a_j \in P_j$ belongs to $\mathcal{C}_{k+1}(\Gamma)$ for some $j \in \overline{L}$. In this case this holds for each $j \in \overline{L}$.

More generally Theorem 2.2.3 says that $C(a^L, P^{\overline{L}})$ with $|L| = k < n$ belongs to $\mathcal{C}_k(\Gamma)$ if and only if each $C(a^K, P^{\overline{K}})$ with $a^{K\setminus L} \in \bigtimes_{K\setminus L} P_i$ belongs to $\mathcal{C}_{k'}(\Gamma)$ with $k' > k$ for some (and then for each) K with $L \subset K, |K| = k'$.

Especially then any $C(a^L, P^{\overline{L}})$ is an implicant of 1_Γ if and only if each $C(a)$ with $a^{\overline{L}} \in \bigtimes_{\overline{L}} P_i$ is a minimplicant of 1_Γ.

Further Lemma 2.1.3 (a) shows how each implicant $C(a^L, P^{\overline{L}}) \in \mathcal{C}_k(\Gamma)$ with $0 \leq k < k' \leq n$ equals, for each K with $L \subset K, |K| = k'$, a sum of implicants $C(a^K, P^{\overline{K}}) \in \mathcal{C}_{k'}(\Gamma)$.

Corollary 2.2.4 yields a generalization of the method of Quine and McClusky to determine the set of all implicants of a Boolean function, see e.g. Wegener (1987). To construct the set $C_k(\Gamma)$ by the set $C_{k+1}(\Gamma)$ ($k = 0, ..., n-1$) with the aid of Corollary 2.2.4 we should have to verify for each cube indicator $C(a^L, P^{\overline{L}})$ with $|L| = k$ whether the condition

$$C(a^{L\cup\{j\}}, P^{\overline{L}\setminus\{j\}}) \in C_{k+1}(\Gamma) \text{ for each } a_j \in P_j$$

holds for some $j \in \overline{L}$ (and then for each $j \in \overline{L}$) or not. Obviously this should be a rather ineffective method. We obtain a more effective method from the following theorem. It shows that we may construct to each element $C(a^K, P^{\overline{K}}) \in C_{k+1}(\Gamma)$ with $1 \in K$ a set of elements of $C_k(\Gamma)$ in such a way that each element of $C_k(\Gamma)$ is obtained exactly once.

2.2.5 Theorem

For any $C(a^K, P^{\overline{K}}) \in C_{k+1}(\Gamma), k = 0, ..., n-1$ and $i \in K$, define

$$N_{i,\Gamma}(a^K, P^{\overline{K}}) := \left\{c_i \in M_i : c_i \geq a_i, C(c^K, P^{\overline{K}}) \in C_{k+1}(\Gamma) \text{ for } c^{K\setminus\{i\}} = a^{K\setminus\{i\}}\right\}.$$

Then

$$C_k(\Gamma) = \dot{\bigcup_{\substack{C(a^K,P^{\overline{K}})\in C_{k+1}(\Gamma)\\ 1\in K}}} \left[\dot{\bigcup_{\substack{i\in K\\ i<\min(k,k\in\overline{K})\\ |N_{i,\Gamma}(a^K,P^{\overline{K}})|>1}}} \left[\dot{\bigcup_{\substack{P_i\subseteq N_{i,\Gamma}(a^K,P^{\overline{K}})\\ a_i\in P_i}}} \left\{C(a^{K\setminus\{i\}}, P^{\overline{K}\cup\{i\}})\right\}\right]\right].$$

Comment

According to Theorem 2.2.5 we obtain $C_k(\Gamma)$ from $C_{k+1}(\Gamma)$ in the following way:

1. We choose out of $C_{k+1}(\Gamma)$ all those elements $C(a^K, P^{\overline{K}})$ with K containing the element 1.
2. To each such $C(a^K, P^{\overline{K}})$ we choose all $i \in K$ which are less than all elements of $\overline{K}$ and for which moreover the set $N_{i,\Gamma}(a^K, P^{\overline{K}})$ does not contain only the element a_i (that means so to speak that $C(a^K, p^{\overline{K}})$ has certain "right neighbours" within $C_{k+1}(\Gamma)$).
3. Finally to each such i we collect all elements $C(a^{K\setminus\{i\}}, P^{\overline{K}\cup\{i\}})$ (out of $C_k(\Gamma)$ of course) for which P_i is a subset of $N_{i,\Gamma}(a^K, P^{\overline{K}})$ containing the smallest element $a_i$of $N_{i,\Gamma}(a^K, P^{\overline{K}})$.

Proof

We write $C'_k(\Gamma)$ for the right hand side. First we justify the notation of $C'_k(\Gamma)$ with mutually disjoint sets.

Clearly for fixed $i \in K$ two elements $C(a^{K\setminus\{i\}}, P^{\overline{K}\cup\{i\}})$ of $C'_k(\Gamma)$ with different sets P'_i and P''_i are different. Further obviously $C(a^{K\setminus\{i\}}, P^{\overline{K}\cup\{i\}})$ and $C(a^{K\setminus\{j\}}, P^{\overline{K}\cup\{j\}})$ for $i, j \in K, i \neq j$ are different.

Finally let $C(a^{K\setminus\{i\}}, P^{\overline{K}\cup\{i\}}) = C(b^{L\setminus\{j\}}, Q^{\overline{L}\cup\{j\}})$ be elements of $C'_k(\Gamma)$. We have to show that $C(a^K, P^{\overline{K}}) = C(b^L, Q^{\overline{L}})$, i.e. $K = L, a^K = b^K, P^{\overline{K}} = Q^{\overline{K}}$. From $\overline{K} \cup \{i\} = \overline{L} \cup \{j\}$ and $i < \min(k : k \in \overline{K}), j < \min(k : k \in L)$ we obtain $i = j = \min(k : k \in \overline{K} \cup \{i\}) = \min(k : k \in \overline{L} \cup \{j\})$ and so $K = L, a^{K\setminus\{i\}} = b^{K\setminus\{i\}}, P^{\overline{K}\cup\{i\}} = Q^{\overline{K}\cup\{i\}}$. Further $a_i = \min(c_i : c_i \in P_i), b_i = \min(c_i : c_i \in Q_i)$, $P_i = Q_i$ implies $a_i = b_i$ and so $a^K = b^K$.

Now we have $C_k(\Gamma) = C'_k(\Gamma)$ to prove. Suppose $C(a^{K\setminus\{i\}}, P^{\overline{K}\cup\{i\}}) \in C'_k(\Gamma)$. Using Lemma 2.1.3 (a) with $L = K \setminus \{i\}$ we may write (see comment to Lemma 2.1.3 (a))

$$C(a^{K\setminus\{i\}}, P^{\overline{K}\cup\{i\}}) = \max_{\substack{b^{K\setminus\{i\}} = a^{K\setminus\{i\}} \\ b_i \in P_i}} C(b^K, P^{\overline{K}}) \quad \text{with } P_i \subseteq N_{i,\Gamma}(a^K, P^{\overline{K}})$$

and so $C(a^{K\setminus\{i\}}, P^{\overline{K}\cup\{i\}}) \leq 1_\Gamma$ by definition of $N_{i,\Gamma}(a^K, P^{\overline{K}})$. Therefore $C(a^{K\setminus\{i\}}, P^{\overline{K}\cup\{i\}}) \in C_k(\Gamma)$ since $|K \setminus \{i\}| = k$.

We have proved $C'_k(\Gamma) \subseteq C_k(\Gamma)$.

To prove $C_k(\Gamma) \subseteq C'_k(\Gamma)$ we suppose $C(b^L, P^{\overline{L}}) \in C_k(\Gamma)$. Define $i_o := \min(j : j \in \overline{L}), K := L \cup \{i_o\}$ so that $1 \in K, i_o \in K, L = K \setminus \{i_o\}, \overline{L} = \overline{K} \cup \{i_o\}$, $i_o < \min(k : k \in \overline{K})$. Let $C(a^K, P^K)$ given by $a^L = b^L$ and $a_{io} := \min(P_{i_o})$. Then $C(b^L, P^{\overline{L}}) = C(a^{K\setminus\{i_o\}}, P^{\overline{K}\cup\{i_o\}})$, further as above

$$C(a^{K\setminus\{i_o\}}, P^{\overline{K}\cup\{i_o\}}) = \max_{\substack{b^{K\setminus\{i_o\}} = a^{K\setminus\{i_o\}} \\ b_{i_o} \in P_{i_o}}} C(b^K, P^{\overline{K}}) \leq 1_\Gamma.$$

Hence $P_{i_o} \subseteq N_{i_o,\Gamma}(a^K, P^{\overline{K}})$ (with $|N_{i_o,\Gamma}(a^K, P^{\overline{K}})| > 1$, since $|P_{i_o}| > 1$), further $C(a^K, P^{\overline{K}}) \in C_{k+1}(\Gamma)$ according to $a_{i_o} \in P_{i_o}$.

Thus we have shown, that $C(b^L, P^{\overline{L}}) = C(a^{K\setminus\{i_o\}}, P^{\overline{K}\cup\{i_o\}})$ satisfies all conditions defining the set $C'_k(\Gamma)$, i.e. that $C(b^L, P^{\overline{L}}) \in C'_k(\Gamma)$, and so $C_k(\Gamma) \subseteq C'_k(\Gamma)$. □

Let us consider any implicant $C(a^K, P^{\overline{K}}) \in C_{k+1}(\Gamma)$ where $1 \in K$. Then for each $i \in K$ with $i < \min(k : k \in \overline{K}), |N_{i,\Gamma}(a^K, P^{\overline{K}})| > 1$ and each P_i with

$a_i \in P_i \subseteq N_{i,\Gamma}(a^K, P^{\overline{K}})$ Theorem 2.2.5 yields exactly one implicant $C(a^{K\setminus\{i\}}, P^{\overline{K}\cup\{i\}}) \in C_k(\Gamma)$. Clearly for fixed $a^{K\setminus\{i\}}, P^{\overline{K}}$ each P_i is a subset of

(a) $\quad N^+_{i,\Gamma}(a^K, P^{\overline{K}}) := \left\{c_i \in M_i : C(c^K, P^{\overline{K}}) \in C_{k+1}(\Gamma) \text{ for } c^{K\setminus\{i\}} = a^{K\setminus\{i\}}\right\}$

which is the same for all implicants $C(b^K, Q^{\overline{K}}) \in C_{k+1}(\Gamma)$ with $b^{K\setminus\{i\}} = a^{K\setminus\{i\}}$, $Q^{\overline{K}} = P^{\overline{K}}$. Therefore it suffices to determine only the set $N^+_{i,\Gamma}(a^K, P^{\overline{K}})$ and to collect all implicants $C(a^{K\setminus\{i\}}, P^{K\cup\{i\}})$ with $P_i \subseteq N^+_{i,\Gamma}(a^K, P^{\overline{K}})$. This leads to the following version of Theorem 2.2.5.

2.2.6 Theorem

For any $C(a^K, P^{\overline{K}}) \in C_{k+1}(\Gamma), k = 0, ..., n-1$ and $i \in K$, define $N^+_{i,\Gamma}(a^K, P^{\overline{K}})$ by (a), further $N'_{i,\Gamma}(a^K, P^{\overline{K}})$ by

$$N'_{i,\Gamma}(a^K, P^{\overline{K}}) := \begin{cases} N^+_{i,\Gamma}(a^K, P^{\overline{K}}) & \text{if } a_i = \min(b_i : b_i \in N^+_{i,\Gamma}(a^K, P^{\overline{K}})), \\ \emptyset & \text{otherwise.} \end{cases}$$

Then

$$C_k(\Gamma) = \dot{\bigcup_{\substack{C(a^K, P^{\overline{K}}) \in C_{k+1}(1_\Gamma) \\ 1 \in K}}} \left[\dot{\bigcup_{\substack{i \in K \\ i < \min(j : j \in \overline{K}) \\ |N'_{i,\Gamma}(a^K, P^{\overline{K}})| > 1}}} \left[\dot{\bigcup_{P_i \subseteq N'_{i,\Gamma}(a^K, P^{\overline{K}})}} \left\{C(a^{K\setminus\{i\}}, P^{\overline{K}\cup\{i\}})\right\}\right]\right].$$

Proof

The proposition follows from Theorem 2.2.5.

For practical computation of $C_k(\Gamma)$ the application of Theorem 2.2.6 could be more efficient than the application of Theorem 2.2.5 (see also the comment to Theorem 2.2.5).

2.3 Prime Implicants

We define so–called prime implicants of 1_Γ by a maximum property. For brevity we sometimes use the notation C instead of $C(P)$ etc.

2.3.1 Definition

An implicant C of 1_Γ is called a *prime* implicant of 1_Γ if there is no implicant C' of 1_Γ with $C \leq C' \neq C$.

To construct the set of all prime implicants of 1_Γ we need the idea of k–maximal implicants.

2.3.2 Definition

An implicant C is called *k–maximal* ($k = 0, ..., n$) if $C \in \mathcal{C}_k(\Gamma)$ and $C \leq C' \neq C$ for no $C' \in \mathcal{C}_k(\Gamma)$.

For $0 < k < n$ this definition means that C is k–maximal exactly if C is of type $C(a^K, P^{\overline{K}})$ with $|K| = k$ and there is no implicant $C(a^K, Q^{\overline{K}}) \in \mathcal{C}_k(\Gamma)$ with $\bigtimes_{\overline{K}} P_i \subset \bigtimes_{\overline{K}} Q_i$. Further $C(P)$ is 0–maximal exactly if $\bigtimes P_i \subset \bigtimes Q_i$ for no other implicant $C(Q) \in \mathcal{C}_o(\Gamma)$. Obviously each minimplicant of 1_Γ is n–maximal.

Now we may characterize the prime implicants of 1_Γ out of $\mathcal{C}_k$ ($k = 0, ..., n$) and thus all prime implicants of 1_Γ by the following statement.

2.3.3 Theorem

(a) Each prime implicant of 1_Γ out of $\mathcal{C}_k(\Gamma)$ is k–maximal, $k = 0, ..., n$.

(b) Each 0–maximal implicant of 1_Γ is a prime implicant of 1_Γ.

(c) A k–maximal implicant $C(a^K, P^{\overline{K}})$ of 1_Γ with $k \neq 0$ is not a prime implicant of 1_Γ if and only if there is another k–maximal implicant $C(b^K, Q^{\overline{K}})$ of 1_Γ with $P_i \subseteq Q_i$ for $i \in \overline{K}$ and $b^{K\setminus\{j\}} = a^{K\setminus\{j\}}, b_j \neq a_j$ for some $j \in K$.

Proof

We only have to prove (c). Thus let $C(a^K, P^{\overline{K}})$ be a k–maximal implicant of 1_Γ with $k \neq 0$ and suppose $C(b^K, Q^{\overline{K}})$ to be such another k–maximal implicant of 1_Γ. Then also $C(b^K, P^{\overline{K}}) \in \mathcal{C}_k(\Gamma)$. From Lemma 2.1.3 (a) with $L = K \setminus \{j\}, P_j = \{a_j, b_j\}$ we obtain

$$C(a^K, P^{\overline{K}}) \leq \max[C(a^K, P^{\overline{K}}), C(b^K, P^{\overline{K}})] = C(a^{K\setminus\{j\}}, P^{\overline{K}\cup\{j\}}) \in \mathcal{C}_{k-1}(\Gamma).$$

Thus $C(a^K, P^{\overline{K}})$ is not a prime implicant.

Now suppose $C(a^K, P^{\overline{K}})$ not to be a prime implicant. Then $C(a^K, P^{\overline{K}}) \leq C(c^L, R^{\overline{L}})$ for some $C(c^L, R^{\overline{L}}) \in \mathcal{C}(\Gamma)$ with $L \neq K$ (since $C(a^K, P^{\overline{K}})$ k–maximal). From Lemma 2.1.3 (c) we obtain

$$L \subset K \ , \ a^L = c^L \quad (\text{if } L \neq \emptyset)$$

$$a^{K\setminus L} \in \bigtimes_{K\setminus L} R_i \ , \ \bigtimes_{\overline{K}} P_i \subseteq \bigtimes_{\overline{K}} R_i \quad (\text{if } K \neq N_n).$$

Now we choose any $j \in K \setminus L$ and define $C(b^K, P^{\overline{K}})$ by $b^{K\setminus\{j\}} = a^{K\setminus\{j\}}$ and any $b_j \in R_j$ with $b_j \neq a_j$. Then

$$b^L = c^L \quad (\text{if } L \neq \emptyset), b^{K\setminus L} \in \bigtimes_{K\setminus L} R_i, \bigtimes_{\overline{K}} P_i \subseteq \bigtimes_{\overline{K}} R_i$$

and so (by Lemma 2.1.3 (c))

$$C(b^K, P^{\overline{K}}) \leq C(c^L, R^{\overline{L}}) \leq 1_\Gamma,$$

i.e. $C(b^K, P^{\overline{K}}) \in \mathcal{C}_k(\Gamma)$ with $b^{K\setminus\{j\}}, b_j \neq a_j$. But then there is also a k–maximal implicant $C(d^K, Q^{\overline{K}})$ of 1_Γ with $C(b^K, P^{\overline{K}}) \leq C(d^K, Q^{\overline{K}})$ and so (due to Lemma 2.1.3 (c) again)

$$d^K = b^K \ , \ \bigtimes_{\overline{K}} P_i \subseteq \bigtimes_{\overline{K}} Q_i.$$

Thus $(b^K, Q^{\overline{K}})$ is a k–maximal implicant of 1_Γ with $\bigtimes_{\overline{K}} P_i \subseteq \bigtimes_{\overline{K}} Q_i$ and $b^{K\setminus\{j\}} = a^{K\setminus\{i\}}, b_j \neq a_j$ for some $j \in K$. □

Theorem 2.3.3 immediately gives a method how to select all prime implicants of 1_Γ out of $\mathcal{C}_k(\Gamma)$ for $k = 0, ..., n$ and so all prime implicants of 1_Γ. To select all prime implicants of 1_Γ out of $\mathcal{C}_k(\Gamma)$ we only need the set of all k–maximal implicants of 1_Γ.

2.4 Representations by Implicants (Prime Implicants)

Now we shall construct representations of 1_Γ as maxima of some of its implicants, especially representations as maxima of prime implicants. Moreover we want to obtain reduced representations which contain no redundant implicants. Finally we look for minimal representations with a minimal number of implicants.

2.4.1 Definition

If $C(P_{(1)}), ..., C(P_{(r)})$ with $P_{(\rho)} := (P_{\rho 1}, ..., P_{\rho n})$ for $\rho \in N_r := \{1, ..., r\}$ are implicants (prime implicants) of 1_Γ and

(2.4.2) $1_\Gamma = \max_{\rho\in N_r} C(P_{(\rho)})$, which is equivalent to $\Gamma = \bigcup_{\rho\in N_r} \bigtimes P_{\rho i} = \bigcup_{\rho\in N_r} \bigcap P^*_{\rho i}$

with $P^*_{\rho i} := \{x \in M : x_i \in P_{\rho i}\}$ (see proof of Lemma 2.1.3) then we call (2.4.2) a *representation of* 1_Γ *by implicants (prime implicants)* (more precisely *by the implicants (prime implicants)* $C(P_{(1)}), ..., C(P_{(r)})$).
The relation (2.4.2) means that $1_\Gamma(x) = 1$ if and only if $C(P_{(\rho)})(x) = 1$ (i.e. $x_1 \in P_{\rho 1}, ..., x_n \in P_{\rho n}$) for at least one $\rho \in N_r$.
We illustrate (2.4.2) by the following parallel circuit of r series circuits (each of them consisting of n parallel circuits as in (2.1.2)):

(2.4.3)

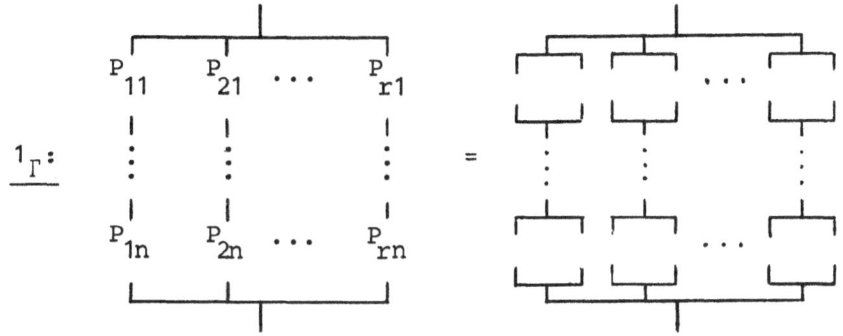

Clearly every indicator 1_Γ has at least one representation by implicants, namely the unique representation by its minimplicants due to (2.2.1). Note that the number of minimplicants of 1_Γ is $|\Gamma|$. Moreover it has at least one representation by prime implicants too: It is easy to see that for each minimplicant $C(a)$ of 1_Γ there exists at least one prime implicant – say C_{pa} of 1_Γ with $C(a) \leq C_{pa}$. Now

$$1_\Gamma = \max_{a \in \Gamma} C_{pa}$$

is a representation of 1_Γ by prime implicants.
In the following we use the simpler notation $C_1, ..., C_r$ instead of $C(P_{(1)}), ..., C(P_{(r)})$ etc.

2.4.4 Definition

Let $C_1, ..., C_r$ be implicants (prime implicants) of 1_Γ with

(2.4.5)
$$1_\Gamma = \max_{\rho \in N_r} C_\rho$$

but

$$(2.4.6) \qquad 1_\Gamma \neq \max_{\rho \in K} C_\rho \quad \text{for each } K \subset N_r.$$

Then we call (2.4.5) a *reduced* representation of 1_Γ by implicants (prime implicants).

Thus a representation of 1_Γ is reduced if and only if it contains no redundant implicants. Clearly (2.4.5) and (2.4.6) imply $1_\Gamma \geq \max_{\rho \in K} C_\rho \neq 1_\Gamma$ (equivalent to $\bigcup_{\rho \in K} \cap P^*_{\rho i} \subset \Gamma$ in the Notation of Definition 2.4.1).

2.4.7 Definition

If (2.4.5) holds and if $s \geq r$ for each representation

$$1_\Gamma = \max_{\rho \in N_s} C'_\rho$$

of 1_Γ by implicants (prime implicants) then we call (2.4.5) a *minimal* representation of 1_Γ by implicants (prime implicants).

A minimal representation is always reduced. Minimal representations are those with the smallest number of implicants.

2.4.8 Definition

A prime implicant C of 1_Γ is called *essential* if $C \in \{C_1, ..., C_r\}$ for each representation (2.4.5) of 1_Γ by prime implicants.

The next statement gives a necessary and sufficient condition that some implicants (prime implicants) of 1_Γ form a representation of 1_Γ according to Definition 2.4.1 (2.4.2).

2.4.9 Theorem

Let $C_1, ..., C_r$ be implicants (prime implicants) of 1_Γ. Then the following are equivalent:

(a)
$$1_\Gamma = \max_{\rho \in N_r} C_\rho$$

(i.e. 1_Γ has a representation by the implicants (prime implicants) $C_1, ..., C_r$).

(b) For each minimplicant $C(a)$ of 1_Γ, there is at least one $C_\rho \in \{C_1, ..., C_r\}$ with $C(a) \leq C_\rho$, which is equivalent to $C_\rho = 1$ for $x = a$.

Proof

Write $C_\rho = C(P_{(\rho)}) = 1_{\times P_{\rho i}}$ (see Definition 2.1.1). Now the following equivalences hold:

$1_\Gamma = \max_{\rho \in N_r} C_\rho \Leftrightarrow \Gamma = \bigcup_{\rho \in N_r} \times P_{\rho i}$

$\Leftrightarrow$ for each $a \in \Gamma$, there is at least one $\rho \in N_r$ with $a \in \times P_{\rho i}$

$\Leftrightarrow$ for each minimplicant $C(a)$ of 1_Γ, there is at least one $C_\rho \in \{C_1, ..., C_r\}$ with $C(a) \leq C_\rho$. □

2.4.10 Corollary

Every indicator 1_Γ has a representation by all its implicants (prime implicants).

2.4.11 Corollary

A prime implicant C of 1_Γ is essential if and only if there exists a minimplicant $C(a)$ of 1_Γ such that C is the only prime implicant of 1_Γ with $C(a) \leq C$.

We wish to gain all reduced (especially all minimal) representations of 1_Γ by implicants (prime implicants). This geometrically corresponds – roughly spoken – to the problem to cover the set Γ by "reduced" (minimal) unions of (maximal) cubes $\times P_i \subseteq \Gamma$.

First we introduce four subsets of the power set $\mathcal{P}(\mathcal{C}(\Gamma))$ (set of all subsets of $\mathcal{C}(\Gamma)$). They characterize those sets of implicants (prime implicants) which constitute representations (reduced representations) of 1_Γ by their maximum.

2.4.12 Definition

Let $\mathcal{C}_p(\Gamma)$ be the set of all prime implicants of 1_Γ.

We define

$$\begin{aligned}
\mathcal{I}(\Gamma) &:= \{B \subseteq \mathcal{C}(\Gamma) : \max_{C \in B} C = 1_\Gamma\},\\
\mathcal{I}_p(\Gamma) &:= \{B \subseteq \mathcal{C}_p(\Gamma) : \max_{C \in B} C = 1_\Gamma\} \subseteq \mathcal{I}(\Gamma),\\
\mathcal{R}(\Gamma) &:= \{B \in \mathcal{I}(\Gamma) \max_{C \in K} C \neq 1_\Gamma \text{ for } K \subset B\} \subseteq \mathcal{I}(\Gamma),\\
\mathcal{R}_p(\Gamma) &:= \{B \in \mathcal{I}_p(\Gamma) : \max_{C \in K} C \neq 1_\Gamma \text{ for } K \subset B\} = \mathcal{R}(\Gamma) \cap \mathcal{I}_p(\Gamma).
\end{aligned}$$

Now it holds

$B \in \mathcal{I}(\Gamma) \Leftrightarrow 1_\Gamma = \max_{C \in B} C$ is a representation of 1_Γ by implicants;

$B \in \mathcal{I}_p(\Gamma) \Leftrightarrow 1_\Gamma = \max_{C \in B} C$ is a representation of 1_Γ by **prime implicants**;
$B \in \mathcal{R}(\Gamma) \Leftrightarrow 1_\Gamma = \max_{C \in B} C$ is a **reduced** representation of 1_Γ by **implicants**;
$B \in \mathcal{R}_p(\Gamma) \Leftrightarrow 1_\Gamma = \max_{C \in B} C$ is a **reduced** representation of 1_Γ by **prime implicants**. Obviously $\mathcal{R}_p(\Gamma)$ is not empty. The following figure shows the order relation of the four sets.

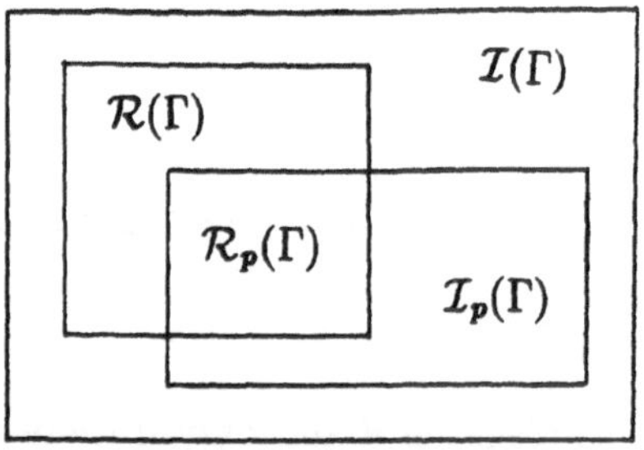

Furthermore we need the following definitions.

2.4.13 Definition

For any finite sets $A_1, ..., A_m$, we define

$$S(A_1, ..., A_m) := \{B \subseteq \bigcup_{N_m} A_\mu : B \cap A_\mu \neq \emptyset \text{ for each } \mu \in N_m\}$$

(set of all subsets of $\bigcup_{N_m} A_\mu$ containing at least one element of each $A_\mu, \mu \in N_m$).

2.4.14 Definition

Let S be any finite set and $\mathcal{T} \subseteq \mathcal{P}(S)$. We define the *reduced* set

$$R(\mathcal{T}) := \{T \in \mathcal{T} : T' \subset T \text{ for no } T' \in \mathcal{T}\}.$$

Now from Theorem 2.4.9 we obtain the following result

2.4.15 Theorem

Let $C(a_{(1)}),...,C(a_{(m)})$ be the minimplicants of 1_Γ (i.e. $\Gamma = \{a_{(1)},...,a_{(m)}\}$). For $\mu \in N_m$, define

$$\mathcal{M}_\mu(\Gamma) := \{C \in \mathcal{C}(\Gamma) : C \geq C(a_{(\mu)})\},$$
$$\mathcal{M}_{p\mu}(\Gamma) := \{C \in \mathcal{C}_p(\Gamma) : C \geq C(a_{(\mu)})\}$$

(thus $\mathcal{C}(\Gamma) = \bigcup_{N_m} \mathcal{M}_\mu(\Gamma), \mathcal{C}_p(\Gamma) = \bigcup_{N_m} \mathcal{M}_{p\mu}(\Gamma)$, see Def. 2.4.12).

Then

$$\mathcal{I}(\Gamma) = \mathcal{S}(\mathcal{M}_1(\Gamma),...,\mathcal{M}_m(\Gamma)),$$
$$\mathcal{I}_p(\Gamma) = \mathcal{S}(\mathcal{M}_{p1}(\Gamma),...,\mathcal{M}_{pm}(\Gamma)),$$
$$\mathcal{R}(\Gamma) = R(\mathcal{I}(\Gamma)),$$
$$\mathcal{R}_p(\Gamma) = R(\mathcal{I}_p(\Gamma)) = \mathcal{R}(\Gamma) \cap \mathcal{I}_p(\Gamma) \text{ (see Definition 2.4.12).}$$

First of all we are interested in the *reduced* representation of 1_Γ, i.e. in the reduced sets $\mathcal{R}(\Gamma)$ and $\mathcal{R}_p(\Gamma)$. We obtain explicite formulas for them by the following general lemma.

2.4.16 Lemma

To any finite sets $A_1,...,A_m (m > 1)$ let $\mathcal{S}(A_1,...,A_m), R(\mathcal{S}(A_1,...,A_m))$ given by Definition 2.4.13 and Definition 2.4.14. For $\mu \in N_m$, define the set $D_\mu(m)$ of "minimal covers" of N_m by

$$D_\mu(m) := \{\{D_1,...,D_\mu\} \subset \mathcal{P}(N_m) : \bigcup_{N_\mu} D_j = N_m, \bigcup_K D_j \subset N_m \text{ for } K \subset N_\mu\}.$$

Then

$$R(\mathcal{S}(A_1,...,A_m)) = \bigcup_{N_m} R_\mu(A_1,...,A_m)$$

with

$$R_\mu(A_1,...,A_m) :=$$
$$\dot{\bigcup_{\{D_1,...,D_\mu\} \in D_\mu(m)}} \left\{ \{\omega_1,...,\omega_\mu\} : \omega_j \in \bigcap_{D_j} A_i \cap \bigcap_{N_m \setminus D_j} \overline{A_i} \text{ for } j \in N_\mu \right\}, \mu \in N_m$$

$(\overline{A_i} := (\bigcup_{N_m} A_j) \setminus A_i)$.

Of course $R_\mu(A_1,...,A_m)$ for $\mu \in N_m$ is the set of all reduced sets out of $\mathcal{S}(A_1,...,A_m)$ of size μ, and Lemma 2.4.16 states that $R(\mathcal{S}(A_1,...,A_m))$ contains to each minimal cover $\{D_1,...,D_\mu\} \in D_\mu(m)$ exactly all subsets $\{\omega_1,...,\omega_\mu\}$ of $\bigcup_{N_m} A_i$ with the property

$$\omega_j \begin{cases} \text{belongs to } A_i \text{ for } i \in D_j, \\ \text{does not belong to } A_i \text{ for } i \notin D_j. \end{cases}$$

We note that $\{\{\omega_1,...,\omega_\mu\} : \omega_j \in \bigcap_{D_j} A_i \cap \bigcap_{N_m \setminus D_j}$ for $j \in N_\mu\}$ does not depend on the succession of $D_1,...,D_\mu$.

Proof.

Write $R := R(\mathcal{S}(A_1,...,A_m)), R' := \dot{\bigcup}_{N_m} R_\mu, R_\mu := R_\mu(A_1,...,A_m), \mu \in N_m$.

We have to prove that $R = R'$.

(a) Suppose $\{\omega_1,...,\omega_\mu\} \in R$. Clearly $\mu \leq m$.
Define $\{D_1,...,D_\mu\}$ by

(2.4.17) $$D_j := \{i \in N_m : \omega_j \in A_i\}\ ,\ j \in N_\mu$$

equivalent to

(2.4.18) $$\omega_j \in (\bigcap_{D_j} A_i) \cap (\bigcap_{N_m \setminus D_j} \overline{A_i})\ ,\ j \in N_\mu.$$

We show that $\{D_1,...,D_\mu\} \in D_\mu(m)$ i.e. $\{\omega_1,...,\omega_\mu\} \in R_\mu \subseteq R'$.
Assume $\bigcup_{N_\mu} D_j \subset N_m$. Then there is a $k \in N_m$ with $k \notin D_j$ (i.e. $k \in N_m \setminus D_j$) and so $\omega_j \in \overline{A_k}$ for $j \in N_\mu$. But then $\{\omega_1,...,\omega_\mu\} \cap A_k = \emptyset$ in contradiction to the supposition $\{\omega_1,...,\omega_\mu\} \in R$. Thus $\bigcup_{N_\mu} D_j = N_m$. Now assume $\bigcup_K D_j = N_m$ for some $K = \{j_1,....,j_k\} \subset N_m$. Then $\{\omega_{j_1},...,\omega_{j_k}\} \subset \{\omega_1,...,\omega_\mu\}$, further $\{\omega_{j_1},...,\omega_{j_k}\} \cap A_i \neq \emptyset$ for $i \in N_m$ due to (2.4.18). But this is again a contradiction to the supposition. Thus $\bigcup_K D_j \subset N_m$ and so $\{D_1,...,D_\mu\} \in D_\mu(m)$.

(b) Now suppose $\{\omega_1,...,\omega_\mu\} \in R'$. Then there is a $\{D_1,...,D_\mu\} \in D_\mu(m)$ with (2.4.18) yielding $\{\omega_1,...,\omega_\mu\} \cap A_i \neq \emptyset$ for $i \in N_m$ and so $\{\omega_1,...,\omega_\mu\} \in \mathcal{S}(A_1,...,.A_m)$. Now consider any $\{\omega_{j_1},...,\omega_{j_k}\} \subset \{\omega_1,...,\omega_\mu\}$. Then $K := \{j_1,...,.j_k\} \subset N_m$ and so $\bigcup_K D_j \subset N_m$. Now in the same way (conf. (a)) it follows $\{\omega_{j_1},...,\omega_{j_k}\} \cap A_j = \emptyset$ for some $j \in N_m$ and so $\{\omega_{j_1},...,\omega_{j_k}\} \notin \mathcal{S}(A_1,...,A_m)$. Thus $\{\omega_1,...,\omega_\mu\} \in R$.

Finally (2.4.18) implies (2.4.17), and so, for each $\{\omega_1, ..., \omega_\mu\} \in R_\mu$, there is exactly one $\{D_1, ..., D_\mu\} \in D(\mu)$ with (2.4.18). Thus we have the proposed representation of $R_\mu, \mu \in N_m$ by **mutually disjoint** sets. □

We illustrate Lemma 2.4.16 by an example.

Example

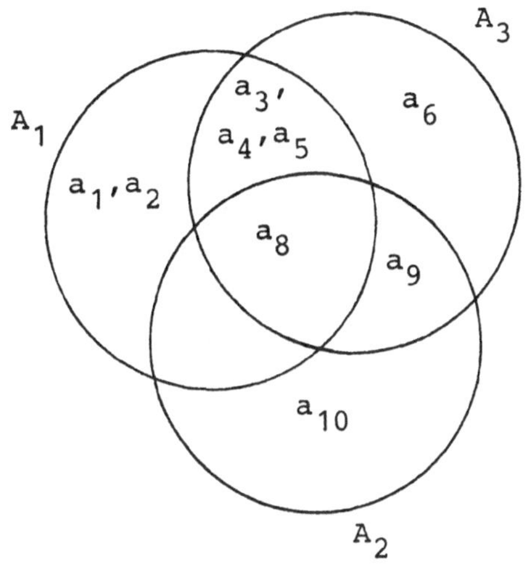

$m = 3,\ N_m = \{1,2,3\}$,

$A_1 = \{a_1, a_2, a_3, a_4, a_5, a_8\}$,

$A_2 = \{a_8, a_9, a_{10}\}$,

$A_3 = \{a_3, a_4, a_5, a_6, a_8, a_9\}$,

$\mathcal{P}(N_3) = \{\{1,\}, \{2\}, \{3\}, \{1,2\}, \{1,3\}, \{2,3\}, \{1,2,3\}\}$,

$D_1(3) = \{\{\{1,2,3\}\}\}$ (one minimal cover)

$\Rightarrow\ R_1(A_1, A_2, A_3) = \{\{\omega_1\} : \omega_1 \in A_1 \cap A_2 \cap A_3 = \{a_8\}\} = \{\{a_8\}\}$,

$D_2(3) = \{\{\{1\},\{2,3\}\}, \{\{2\},\{1,3\}\}, \{\{3\},\{1,2\}\}, \{\{1,2\},\{1,3\}\},$
$\{\{1,2\},\{2,3\}\}, \{\{1,3\},\{2,3\}\}\}$ (6 minimal covers),

where $A_1 \cap A_2 \cap \overline{A}_3 = \emptyset$

$\Rightarrow\ R_2(A_1, A_2, A_3) = \{\{\omega_1, \omega_2\} : \omega_1 \in A_1 \cap \overline{A}_2 \cap \overline{A}_3 = \{a_1, a_2\},$
$\omega_2 \in \overline{A}_1 \cap A_2 \cap A_3 = \{a_9\}\}$

$$\cup\{\{\omega_1,\omega_2\}:\omega_1\in\overline{A}_1\cap A_2\cap\overline{A}_3=\{a_{10}\},\omega_2\in A_1\cap\overline{A}_2\cap A_3=\{a_3,a_4,a_5\}\}$$
$$\cup\{\{\omega_1,\omega_2\}:\omega_1\in A_1\cap\overline{A}_2\cap A_3=\{a_3,a_4,a_5\},\omega_2\in\overline{A}_1\cap A_2\cap A_3=\{a_9\}\}$$
$$=\{\{a_1,a_9\},\{a_2,a_9\}\}$$
$$\cup\{\{a_3,a_{10}\},\{a_4,a_{10}\},\{a_5,a_{10}\}\}$$
$$\cup\{\{a_3,a_9\},\{a_4,a_9\},\{a_5,a_9\}\},$$

$D_3(3)=\{\{\{1\},\{2\},\{3\}\}\}$ (one minimal cover)

$$\Rightarrow\quad R_3(A_1,A_2,A_3)=\{\{\omega_1,\omega_2,\omega_3\}:\omega\in A_1\cap\overline{A}_2\cap\overline{A}_3=\{a_1,a_2\},$$
$$\omega_2\in\overline{A}_1\cap A_2\cap\overline{A}_3=\{a_{1o}\},\omega_3\in\overline{A}_1\cap\overline{A}_2\cap A_3=\{a_6\}\}=\{\{a_1,a_6,a_{10}\},$$
$$\{a_2,a_6,a_{10}\}\}.$$

Thus

$$R(\mathcal{S}(A_1,A_2,A_3))=\{\{a_8\},\{a_1,a_9\},\{a_2,a_9\},\{a_3,a_9\},\{a_4,a_9\},\{a_5,a_9\}$$
$$\{a_3,a_{10}\},\{a_4,a_{10}\},\{a_5,a_{10}\},\{a_1,a_6,a_{10}\},\{a_2,a_6,a_{10}\}\}.$$

In the preceding we deduced a method to find all representations as well as all reduced representations of 1_Γ by implicants (prime implicants). In Lemma 2.4.16 we have the fundamental result yielding the "reduced system" ($R(\mathcal{S}(A_1,...,A_m))$ to a given system $\mathcal{S}(A_1,...,.A_n)$. But practical applications of Lemma 2.4.16 may be painful, since the number $|D_\mu(m)|$ of minimal covers of N_m by μ of its subsets becomes large even for relatively small m and μ (e.g. 3410 for $m=6$, $\mu=3$). We also remark that for some $\{D_1,...,D_\mu\}$ the set of all $\{\omega_1,...,\omega_\mu\}$ with (2.4.18) may be empty. Therefore we need more efficient methods to determine the reduced system $R(\mathcal{S}(A_1,...,A_m))$ to any given sets $A_1,...,A_m$. We shall give such reduction methods in Chapter 4. It will become apparent that they may be applied in the same way to find redûced representations of binary functions by implicates which will be investigated in the next chapter.

Chapter 3

Representations of Binary Functions by Implicates

In this chapter we consider representations of a binary function 1_Γ by minima (products) of implicates of them defined by anticube indicators, i.e. indicators of complements of cubes.

3.1 Anticube Indicators

3.1.1 Definition

For any $P_1 \subseteq M_1, ..., P_n \subseteq M_n$, let again $P_1^* \subseteq M, ..., P_n^* \subseteq M$ be given by (2.1.4) (see proof of Lemma 2.1.3).

If $\emptyset \neq \bigcup P_i^* \subset M$ we call $\bigcup P_i^*$ an *anticube* and $F(P) := F(P_1, ..., P_n : M \to \{0,1\}$, defined by

$$F(P_1, ..., P_n) := 1_{\bigcup P_i^*}$$

an *anticube indicator.*

If $P_i = \overline{\{a_i\}} := M_i \setminus \{a_i\}$ (and so $|P_i| = M_i - 1$) for $i = 1, ..., n$, then we write $F(\overline{a})$ and call it a *maxterm.*

Since $\bigcup \overline{\{a_i\}}^{\,*} = \overline{\{a\}}$ (with $a \in M$), it holds

$$F(\overline{a}) = 1_{\overline{\{a\}}}.$$

If $\emptyset \neq K \subset N_n$ and $P_i = \overline{\{a_i\}}$ for $i \in K, |P_i| < M_i - 1$ for $i \in \overline{K}$ then we write $F(\overline{a}^K, P^{\overline{K}})$ instead of $F(P)$. Further we use $F(\overline{a}^\phi, P^{N_n}) := F(P)$ and $F(\overline{a}^{N_n}, P^\phi) := F(\overline{a})$.

It is easy to see that the anticube $\bigcup P_i^*$ is the complement to the cube $\times \overline{P}_i$ (with $\overline{P}_i := M_i \setminus P_i$): From (2.1.4) we obtain $\overline{P_i^*}(:= M \setminus P_i^*) = \overline{P}_i^*$. Now de Morgan's rule yields

$$\bigcup P_i^* = \overline{\bigcap \overline{P_i^*}} = (:= M \setminus \bigcap \overline{P_i^*}) = \overline{\bigcap \overline{P}_i^*} = \overline{\times \overline{P}_i}.$$

From Definition 2.1.1 and Definition 3.1.1 now we obtain

$$F(P) = 1 - C(\overline{P}) \tag{3.1.2}$$

with $C(\overline{P}) := C(\overline{P}_1, ..., \overline{P}_n)$. Further from (2.1.4) we see that an anticube indicator $F(P)$ takes 1 if and only if $x_i \in P_i$ for at least one $i \in N_n$. This suggests to illustrate $F(P)$ by a parallel circuit of n parallel circuits:

(3.1.3)

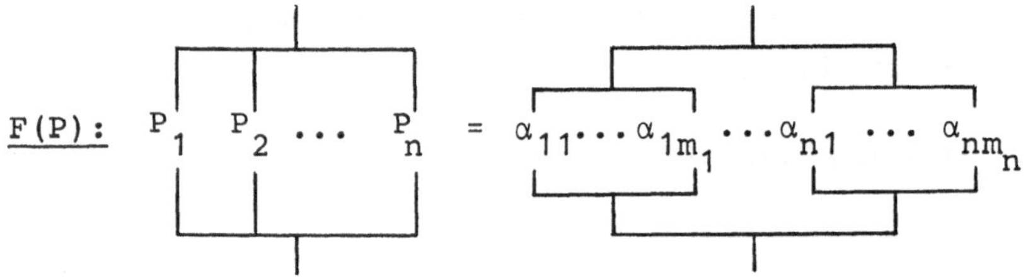

($F(P)(x) = 1$ if and only if $x_1 \in P_1$ or $X_2 \in P_2$ or ... or $x_n \in P_n$).

The next result is the dual version of Lemma 2.1.3. It may be deduced from Lemma 2.1.3 and (3.1.2).

3.1.4 Lemma

(a) If $L \subset K$ then

$$F(\overline{a}^L, P^{\overline{L}}) = \prod_{a^{K\setminus L} \in \bigtimes_{K\setminus L} \overline{P}_i} F(\overline{a}^K, P^{\overline{K}}) = \min_{a^{K\setminus L} \in \bigtimes_{K\setminus L} \overline{P}_i} F\overline{a}^K, P^{\overline{K}})$$

(see also the comment to Lemma 2.1.3 (a)).

(b) The following are equivalent:
(α) $F(P) \leq F(Q)$.
(β) $\bigtimes P_i \subseteq \bigtimes Q_i$.

(c) The following are equivalent:
(α) $F(\overline{a}^K, P^{\overline{K}}) \leq F(\overline{b}^L, Q^{\overline{L}})$.
(β) $K \subseteq L, a^K = b^K$ (if $K \neq \emptyset$), $b^{L\setminus K} \in \bigtimes_{L\setminus K} \overline{P}_i$ (if $K \subset L$), $\bigtimes_{\overline{L}} P_i \subseteq \bigtimes_{\overline{L}} Q_i$ (if $L \neq N_n$).

3.2 Implicates

Let again $1_\Gamma : M \to \{0,1\}$ be any binary function ($\emptyset \neq \Gamma \subset M$).

First we state the unique representation of 1_Γ as a product respectively minimum of maxterms $F(\overline{a})$ given by

$$1_\Gamma = \prod_{a\in\overline{\Gamma}} F(\overline{a}) = \min_{a\in\overline{\Gamma}} F(\overline{a}) \tag{3.2.1}$$

with

$$a \in \overline{\Gamma} \Leftrightarrow C(a) \leq 1_{\overline{\Gamma}} \Leftrightarrow 1_\Gamma \leq F(\overline{a}) \Leftrightarrow \Gamma \subseteq \overline{\{a\}}. \tag{3.2.2}$$

Proof

Using (2.2.1) for $1_{\overline{\Gamma}}$ and (3.1.2) with $P_i = \{a_i\}$ for $i \in N_n$ we obtain

$$1_\Gamma = 1 - 1_{\overline{\Gamma}} = 1 - \max_{a\in\overline{\Gamma}} C(a) = \min_{a\in\overline{\Gamma}}(1 - C(a)) = \min_{a\in\overline{\Gamma}} F(\overline{a}).$$

Further

$$\Gamma \subseteq \overline{\{a\}} \Leftrightarrow a \in \overline{\Gamma} \Leftrightarrow C(a) \leq 1_{\overline{\Gamma}} \Leftrightarrow 1 - F(\overline{a}) \leq 1 - 1_\Gamma \Leftrightarrow 1_\Gamma \leq F(\overline{a}).$$

The uniqueness of the representation is evident. □

We may regard (3.2.1) as the dual counterpart to the representation (2.2.1) of 1_Γ by minterms (more precisely by its minimplicants).

Now we define implicates of 1_Γ as special anticube indicators.

3.2.3 Definition

An anticube indicator $F(P)$ is called an *implicate* of 1_Γ if and only if $F(P) \geq 1_\Gamma$, equivalent to $\Gamma \subseteq \bigcup P_i^*$. The implicates $F(\overline{a})$ of 1_Γ are called the *maximplicates* of 1_Γ.

According to (3.2.2) a maxterm $F(\overline{a})$ is a maximplicate of 1_Γ if $C(a)$ is a minimplicant of $1_{\overline{\Gamma}}$, which is equivalent to $a \in \overline{\Gamma}$ respectively $\Gamma \subseteq \overline{\{a\}}$, and 1_Γ equals the product (the minimum) of all its maximplicates (see 3.2.1)). Further an anticube indicator $F(P)$ is an implicate of 1_Γ if and only if $1_\Gamma(x) = 1$ always implies $F(P)(x) = 1$ (i.e. $x_1 \in P_1$ or ... or $x_n \in P_n$).

From (3.1.2) and $1 - 1_\Gamma = 1_{\overline{\Gamma}}$ immediately we obtain

$$(3.2.4) \qquad F(P) \geq 1_\Gamma \Longleftrightarrow C(\overline{P}) \leq 1_{\overline{\Gamma}}.$$

This immediately yields the following result.

3.2.5 Theorem

The following are equivalent:

(a) The anticube indicator $F(P)$ is an implicate of 1_Γ.

(b) The cube indicator $C(\overline{P})$ is an implicant of $1_{\overline{\Gamma}}$.

Theorem 3.2.5 yields a first possibility to gain the set of all implicates of 1_Γ from the set of all implicants of $1_{\overline{\Gamma}}$.

The second way to construct the set of all implicates of 1_Γ is analogous to that of Chapter 2 to obtain the set $\mathcal{C}(\Gamma)$ of all implicants of 1_Γ.

First we state the implicate–versions of Theorem 2.2.3 and Corollary 2.2.4. They may be deduced from Lemma 3.1.5 or from Theorem 2.2.3 and Corollary 2.2.4 by use of (3.1.2). We give them without proof.

Denote by $\mathcal{F}(\Gamma)$ the set of all implicates of 1_Γ.
For $k = 0, ..., n$ define

$$\mathcal{F}_k(\Gamma) := \{F(\overline{a}^K, P^{\overline{K}}) \in \mathcal{F}(\Gamma) : |K| = k\}$$

($\mathcal{F}_0(\Gamma) = \{F(P) \in \mathcal{F}(\Gamma) : |P_i| < M_i - 1 \text{ for } i \in N_n\}$,
$\mathcal{F}_n(\Gamma) = \{F(\overline{a}) : \mathcal{F}(\overline{a}) \in \mathcal{F}(\Gamma)\}, \mathcal{F}_n(\Gamma)$ is the set of all maximplicates of 1_Γ),
thus obviously

$$\mathcal{F}(\Gamma) = \bigcup_0^n \mathcal{F}_k(\Gamma).$$

3.2.6 Theorem (conf. Theorem 2.2.3)

Let $L \subset N_n$ with $|L| = k$. Then (for any anticube indicator $F(\overline{a}^L, P^{\overline{L}})$) the following are equivalent:

(a) $F(\overline{a}^L, P^{\overline{L}}) \in \mathcal{F}_k(\Gamma)$.

(b) For some K with $L \subset K \subseteq N_n, |K| = k'$ it holds:
$F(\overline{a}^K, P^{\overline{K}}) \in \mathcal{F}_{k'}(\Gamma)$ for each $a^{K \setminus L} \in \bigtimes_{K \setminus L} \overline{P_i}$.

(c) For each K with $L \subset K \subseteq N_n, |K| = k'$ it holds:
$F(\overline{a}^K, P^{\overline{K}}) \in \mathcal{F}_{k'}(\Gamma)$ for each $a^{K \setminus L} \in \bigtimes_{K \setminus L} \overline{P_i}$.

3.2.7 Corollary

Let $L \subset N_n$ with $|L| = k$. Then (for any anticube indicator $F(\overline{a}^L, P^{\overline{L}})$ the following are equivalent:

(a) $F(\overline{a}^L, P^{\overline{L}}) \in \mathcal{F}_k(\Gamma)$.

(b) For some $j \in \overline{L}$ holds:
$F(\overline{a}^{L \cup \{j\}}, P^{\overline{L} \setminus \{j\}}) \in \mathcal{F}_{k+1}(\Gamma)$ for each $a_j \in \overline{P_j}$.

(c) For each $j \in \overline{L}$ holds:
$F(\overline{a}^{L \cup \{j\}}, P^{\overline{L} \setminus \{j\}}) \in \mathcal{F}_{k+1}(\Gamma)$ for each $a_j \in \overline{P_j}$.

The following remarks correspond to those on Theorem 2.2.3. and Corollary 2.2.4. Corollary 3.2.7 in principle yields a method to gain stepwise $\mathcal{F}_{n-1}(\Gamma)$, $\mathcal{F}_{n-2}(\Gamma), ..., \mathcal{F}_0(\Gamma)$ beginning at the set $\mathcal{F}_n(\Gamma)$ of all maximplicates $F(\overline{a})$ of 1_Γ. Evidently if any $\mathcal{F}_k(\Gamma)$ is empty then also $\mathcal{F}_{k-1}(\Gamma)$. We note that according to Corollary 3.2.7 any $F(\overline{a}^L, P^{\overline{L}})$ belongs to $\mathcal{F}_k(\Gamma)$ if and only if each $F(\overline{a}^{L \cup \{j\}}, P^{\overline{L} \setminus \{j\}})$ with $a_j \in P_j$ belongs to $\mathcal{F}_{k+1}(\Gamma)$ for *some* $j \in \overline{L}$. In this case this holds for *each* $j \in \overline{L}$.

More generally Theorem 3.2.6 says that $F(\overline{a}^L, P^{\overline{L}})$ with $|L| = k < n$ belongs to $\mathcal{F}_k(\Gamma)$ if and only if each $F(\overline{a}^K, P^{\overline{K}})$ with $a^{K \setminus L} \in \bigtimes_{K \setminus L} \overline{P_i}$ belongs to $F_{k'}$ with $k' > k$ for some (and then for each) K with $L \subset K, |K| = k'$.

Especially then any $F(\overline{a}^L, \overline{P}^L)$ belongs to $\mathcal{F}(\Gamma)$ (is an implicate of 1_Γ) if and only if each $F(\overline{a})$ with $a^{\overline{L}} \in \bigtimes_{\overline{L}} \overline{P}_i$ is a maximplicate of 1_Γ.

Further Lemma 3.1.4 (a) shows how each implicate $F(\overline{a}^L, P^{\overline{L}}) \in \mathcal{F}_k(\Gamma)$ with $0 \le k < k' \le n$ equals (to any K with $L \subset K \subseteq N_n, |K| = k'$) a product of implicates $F(\overline{a}^K, P^{\overline{K}}) \in \mathcal{F}_{k'}(\Gamma)$.

As in the case of implicants we see that the construction of the set $\mathcal{F}(\Gamma)$ of all implicates of 1_Γ by the aid of Corollary 3.2.7 is not very efficient (see comment to Corollary 2.2.4). Therefore we state the following implicate analogues to Theorem 2.2.5 and Theorem 2.2.6.

3.2.8 Theorem (conf. Theorem 2.2.5)

For any $F(\overline{a}^K, P^{\overline{K}}) \in \mathcal{F}_{k+1}(\Gamma), k = 0, ..., n-1$ and $i \in K$ define

$$N^*_{i,\Gamma}(\overline{a}^K, P^{\overline{K}}) := \left\{ c_i \in M_i : c_i \ge a_i, F(\overline{c}^K, P^{\overline{K}}) \in \mathcal{F}_{k+1}(\Gamma) \text{ for } c^{K \setminus \{i\}} = a^{K \setminus \{i\}} \right\}.$$

Then

$$\mathcal{F}_k(\Gamma) = \dot{\bigcup_{\substack{F(\overline{a}^K, P^{\overline{K}}) \in \mathcal{F}_{k+1}(\Gamma) \\ 1 \in K}}} \left[\dot{\bigcup_{\substack{i \in K \\ i < \min(k, k \in \overline{K}) \\ |N^*_{i,\Gamma}(\overline{a}^K, P^{\overline{K}})| > 1}}} \left[\dot{\bigcup_{\substack{\overline{P_i} \subseteq N^*_{i,\Gamma}(\overline{a}^K, P^{\overline{K}}) \\ P_i \subset \overline{\{a_i\}}}}} \{F(\overline{a}^{K \setminus \{i\}}, P^{\overline{K} \cup \{i\}})\} \right]\right].$$

Proof

Write $\mathcal{F}'_k(\Gamma)$ for the right hand side. Then
$F(\overline{b}^L, Q^{\overline{L}}) \in \mathcal{F}_k(\Gamma) \Leftrightarrow C(b^L, \overline{Q}^{\overline{L}}) \in \mathcal{C}_k(\overline{\Gamma}) \Leftrightarrow C(b^L, \overline{Q}^{\overline{L}}) = C(a^{K\setminus\{i\}}, P^{\overline{K}\cup\{i\}})$
with $P_i \subseteq N_{i,\overline{\Gamma}}(a^K, p^{\overline{K}}), a_i \in P_i, i \in K, i < \min(k, k \in \overline{K})$,
$|N_{i,\overline{\Gamma}}(a^K, P^{\overline{K}})| > 1, C(a^K, P^{\overline{K}}) \in \mathcal{C}_{k+1}(\overline{\Gamma}), 1 \in K$ (according to Theorem 2.2.5)
$\Leftrightarrow F(\overline{b}, Q^{\overline{L}}) = F(\overline{a}^{K\setminus\{i\}} \overline{P}^{\overline{K}\cup\{j\}})$
with $P_i \subseteq N^*_{i,\Gamma}(\overline{a}^K, \overline{P}^{\overline{K}}), \overline{P}_i \subset \overline{\{a_i\}}, i \in K, i < \min(k, k \in \overline{K})$,
$|N^*_{i,\Gamma}(\overline{a}^K, \overline{P}^{\overline{K}})| > 1, F(\overline{a}^K, \overline{P}^{\overline{K}}) \in \mathcal{F}_{k+1}(\Gamma), 1 \in K$
$\Leftrightarrow F(\overline{b}, Q^{\overline{L}}) = F(\overline{a}^{K\setminus\{i\}}, P^{\overline{K}\cup\{i\}})$
with $\overline{P}_i \subseteq N^*_{i,\Gamma}(\overline{a}^K, P^{\overline{K}}), P_i \subset \overline{\{a_i\}}, i \in K, i < \min(k, k \in \overline{K})$,
$|N^*_{i,\Gamma}(\overline{a}^K, P^{\overline{K}})| > 1, F(\overline{a}^K, P^{\overline{K}}) \in \mathcal{F}_{k+1}(\Gamma), 1 \in K$ (by change of P and $\overline{P}$)
$\Leftrightarrow F(\overline{b}, Q^{\overline{L}}) \in \mathcal{F}'_k(\Gamma)$.

The representation of $\mathcal{F}_k(\Gamma)$ with mutually disjoint sets is justified by the fact, that $\mathcal{C}_k(\overline{\Gamma})$ has such a representation according to Theorem 2.2.5. □

The following version of theorem 3.2.8 is the implicate analogue to Theorem 2.2.6. It may be useful to practical computation of $\mathcal{F}_k(\Gamma)$ (see also the consideration before Theorem 2.2.6).

3.2.9 Theorem (conf. Theorem 2.2.6)

For any $F(\overline{a}^K, P^{\overline{K}}) \in \mathcal{F}_{k+1}(\Gamma), k = 0, ..., n-1$ and $i \in K$, define

$$N^{*+}_{i,\Gamma}(\overline{a}^K, P^{\overline{K}}) := \left\{ c_i \in M_i : F(\overline{c}^K, P^{\overline{K}}) \in \mathcal{F}_{k+1}(\Gamma) \text{ for } c^{K\setminus\{i\}} = a^{K\setminus\{i\}} \right\},$$

further

$$N^{*\prime}_{i,\Gamma}(\overline{a}^K, P^{\overline{K}}) := \begin{cases} N^{*+}_{i,\Gamma}(\overline{a}^K, P^{\overline{K}}) & \text{if } a_i = \min(b_i : b_i \in N^{*+}_{i,\Gamma}(\overline{a}^K, P^{\overline{K}}) \\ \emptyset & \text{otherwise.} \end{cases}$$

Then

$$\mathcal{F}_k(\Gamma) =$$

$$\bigcup_{\substack{F(\overline{a}^K, P^{\overline{K}}) \in \mathcal{F}_{k+1}(\Gamma) \\ 1 \in K}} \left[\bigcup_{\substack{i \in K \\ i < \min(k, k \in \overline{K}) \\ |(N^{*\prime}_{i,\Gamma}(\overline{a}, P^{\overline{K}})| > 1}} \left[\bigcup_{\overline{P_i} \subseteq N^{*\prime}_{i,\Gamma}(\overline{a}^K, P^{\overline{K}})} \left\{ F(\overline{a}^{K \setminus \{i\}}, P^{\overline{K} \cup \{i\}}) \right\} \right] \right].$$

3.3 Prime Implicates

On the analogy of prime implicants in Chapter 2 now we define prime implicates of 1_Γ by a minimum property. In the following for brevity sometimes we use the notation F instead of $F(P)$ etc.

3.3.1 Definition

An implicate F of 1_Γ is called a *prime* implicate of 1_Γ if there is no implicate F' of 1_Γ with $F \neq F' \leq F$.

First from Theorem 3.2.5 and Definition 3.3.1 we obtain the prime implicate version of Theorem 3.2.5.

3.3.2 Theorem

The following are equivalent:

(a) The anticube indicator $F(P)$ is a prime implicate of 1_Γ.

(b) The cube indicator $C(\overline{P})$ is a prime implicant of $1_{\overline{\Gamma}}$.

This yields the possibility to gain the prime implicates of 1_Γ from the prime implicants of $1_{\overline{\Gamma}}$.

On the other hand we may choose the prime implicates of 1_Γ out of the set of implicants with the aid of the "k–minimal" implicates of 1_Γ corresponding to the k–maximal implicants of 1_Γ in Chapter 2.

3.3.3 Definition

An implicate F of 1_Γ is called *k-minimal* ($k = 0, ..., n$) if and only if $F \in \mathcal{F}_k(\Gamma)$ and $F \neq F' \leq F$ for no $F' \in \mathcal{F}_k(\Gamma)$.

Now we state the implicate version of Theorem 2.3.3.

3.3.4 Theorem (conf. Theorem 2.3.3)

(a) Each prime implicate of 1_Γ out of $\mathcal{F}_k(\Gamma)$ is k–minimal.

(b) Each 0–minimal implicate of 1_Γ is a prime implicate of 1_Γ.

(c) A k–minimal implicate $F(\overline{a}^K, P^{\overline{K}})$ of 1_Γ with $k \neq 0$ is not a prime implicate of 1_Γ if and only if there is another k–minimal implicate $F(\overline{b}^K, Q^{\overline{K}})$ of 1_Γ with $Q_i \subseteq P_i$ for $i \in \overline{K}$ and $b^{K\setminus\{j\}} = a^{K\setminus\{j\}}, b_j \neq a_j$ for some $j \in K$.

Proof

We only have to prove (c). But this follows from Theorem 2.3.3 by the fact that any $F(\overline{a}^K, P^{\overline{K}})$ is a k–minimal implicate respectively a prime implicate of 1_Γ if and only if $C(a^K, \overline{P}^{\overline{K}})$ is a k–maximal implicant respectively a prime implicant of $1_{\overline{\Gamma}}$. □

Theorem 3.3.3 now yields the set of all prime implicates of 1_Γ out of $\mathcal{F}_k$ as a subset of all k–minimal implicates of 1_Γ.

3.4 Representations by Implicates (Prime Implicates)

Representations of 1_Γ as minima (products) of some of its implicates (prime implicates) are the dual counterparts to the representations of 1_Γ as maxima of some of its implicants (prime implicants).

3.4.1 Definition

If $F(P_{(1)}), ..., F(P_{(r)}))$ with $P_{(\rho)} = (P_{\rho 1}, ..., P_{\rho n})$ for $\rho \in N_r$ are implicates (prime implicates) of 1_Γ and
(3.4.2)

$$1_\Gamma = \min_{\rho \in N_r} F(P_{(\rho)}) = \prod_{\rho \in N_r} F(P_{(\rho)}), \text{ which is equivalent to } \Gamma = \bigcap_{\rho \in N_r} \bigcup P^*_{\rho i},$$

then we call (3.4.2) *a representation of* 1_Γ *by implicates (prime implicates)* (more precisely: by the implicates (prime implicates) $F(P_{(1)}), ..., F(P_{(r)}))$.

Obviously this is a counterpart to the representation (2.4.2) of 1_Γ by implicants. The relation (3.4.2) means that $1_\Gamma(x) = 1$ if and only if $F(P_{(\rho)})(x) = 1$ (i.e. $x_1 \in P_{\rho 1}$ or ... or $x_n \in P_{\rho n}$) for each $\rho \in N_r$. The corresponding illustration is a series circuit of r parallel circuits (each of them consisting of n parallel circuits as in (3.1.3):

(3.4.3) $\underline{1_\Gamma:}$

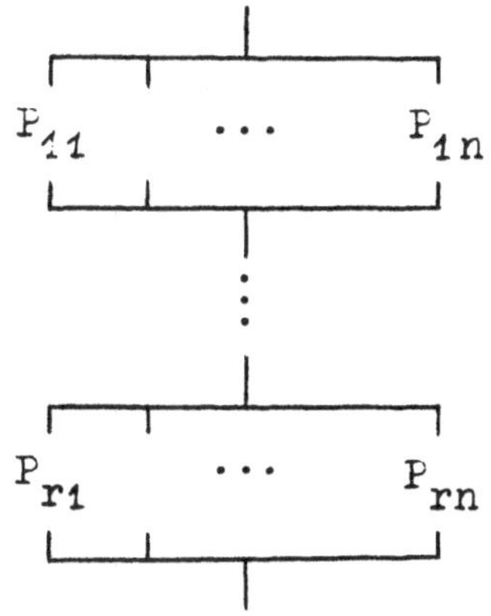

Every indicator 1_Γ has at least the unique representation (3.2.1) by its maximplicates. Note that the number of maximplicates of 1_Γ is $|\overline{\Gamma}|$ with $|\overline{\Gamma}| + |\Gamma| = |M| = \prod |M_i| = \prod (k_i + 1)$ where $|\Gamma|$ is the number of minimplicants of 1_Γ.

Moreover 1_Γ has at least one representation by prime implicates too: Obviously for each maximplicate $F(\overline{a})$ of 1_Γ, there exists at least one prime implicate – say F_{pa} – of 1_Γ with $F_{pa} \leq F(\overline{a})$.

Now

$$1_\Gamma = \min_{a \in \overline{\Gamma}} F_{pa}$$

is a representation of 1_Γ by prime implicates.

The next definitions are analogues to the definitions 2.4.4, 2.4.7, 2.4.8. We use the simpler notation $F_1, ..., F_r$ instead of $F(P_{(1)}), ..., F(P_{(r)})$ etc. The term min always may be replaced by $\prod$.

3.4.4 Definition

Let $F_1, ..., F_r$ be implicates (prime implicates) of 1_Γ with

(3.4.5) $$1_\Gamma = \min_{\rho \in N_r} F_\rho$$

but

(3.4.6) $$1_\Gamma \neq \min_{\rho \in K} F_\rho \quad \text{for each} \quad K \subset N_r.$$

Then we call (3.4.5) a *reduced* representation of 1_Γ by implicates (prime implicates). Thus a representation of 1_Γ is reduced if and only if it contains no redundant implicates. Clearly (3.4.5) and (3.4.6) imply $1_\Gamma \leq \min_{\rho \in K} F_\rho \neq 1_\Gamma$, equivalent to $\Gamma \subset \bigcap_{\rho \in K} \bigcup P^*_{\rho i}$ (see Definition 3.4.1).

3.4.7 Definition

If (3.4.5) holds and $s \geq r$ for each representation

$$1_\Gamma = \min_{\rho \in N_s} F'_\rho$$

of 1_Γ by implicates (prime implicates) then we call (3.4.5) a *minimal* representation of 1_Γ by implicates (prime implicates).

A minimal representation is always reduced. Minimal representations are those with the smallest number of implicates.

3.4.8 Definition

A prime implicate F of 1_Γ is called *essential* if

$$F \in \{F_1, ..., F_r\}$$

for each representation (3.4.5) of 1_Γ by prime implicates.

The next statement gives

a) a generalization of Theorem 3.2.5 and Theorem 3.3.2,
b) the connection between the representations of 1_Γ by implicates (prime implicates) and the representations of $1_{\overline{\Gamma}}$ by implicants (prime implicants).

3.4.9 Theorem

(a) The following are equivalent:
   (α) $F(P)$ is an implicate (maximplicate, prime implicate, essential prime implicate) of 1_Γ.
   (β) $C(\overline{P})$ is an implicant (minimplicant, prime implicant, essential prime implicant) of $1_{\overline{\Gamma}}$.

(b) The following are equivalent:

(α) The indicator 1_Γ has the (reduced, minimal) representation

$$1_\Gamma = \min_{\rho \in N_r} F(P_{(\rho)})$$

by implicates (prime implicates).

(β) The indicator $1_{\overline{\Gamma}}$ has the (reduced, minimal) representation

$$1_{\overline{\Gamma}} = \max_{\rho \in N_r} C(\overline{P}_{(\rho)})$$

by implicants (prime implicants).

Proof

From the identity

$$\min_{\rho \in N_r} F(P_{(\rho)}) = 1 - \max_{\rho \in N_r}(1 - F(P_{(\rho)}))$$

and (3.1.2) we obtain

$$\min_{\rho \in N_r} F(P_{(\rho)}) = 1 - \max_{\rho \in N_r} C(\overline{P}_{(\rho)})$$

and so the equivalence

$$1_\Gamma = \min_{\rho \in N_r} F(P_{(\rho)}) \quad \Leftrightarrow \quad 1_{\overline{\Gamma}} = \max_{\rho \in N_r} C(\overline{P}_{(\rho)}).$$

This is the proposition (b) for representations by implicates, respectively implicants. The remaining statements of (a) and (b) are consequences of the several definitions. □

From Theorem 3.4.9 we learn that we gain all relevant results about (reduced, minimal) representations of 1_Γ by (prime) implicates from the results about the corresponding representations of $1_{\overline{\Gamma}}$ by implicants given in section 2.

The following results concerning implicates (prime implicates) as well as (reduced) representations of 1_Γ by them may be deduced from the corresponding results of section 2 by (3.1.2) and Theorem 3.4.9. Therefore we give them without proofs.

3.4.10 Theorem (see Theorem 2.4.9)

Let $F_1, ..., F_r$ be implicates (prime implicates) of 1_Γ. Then the following are equivalent:

(a) $1_\Gamma = \min_{\rho \in N_r} F_\rho$.

(b) For each maximplicate $F(\overline{a})$ of 1_Γ, there is at least one $F_\rho \in \{F_1, ..., F_r\}$ with $F_\rho \leq F(\overline{a})$.

3.4.11 Corollary (see Corollary 2.4.10)

Every indicator has a representation by all its implicates (prime implicates).

3.4.12 Corollary (see Corollary 2.4.11)

A prime implicate F of 1_Γ is essential if and only if there exists a maximplicate $F(\overline{a})$ of 1_Γ such that F is the only prime implicate of 1_Γ with $F \leq F(\overline{a})$.

The following statements are the complete implicate analogues to the results concerning the representations of 1_Γ by implicants given in chapter 2. Now a reduced (minimal) representation of 1_Γ by implicates (prime implicates) geometrically corresponds to a representation of the set Γ as "reduced" (minimal) intersections of (minimal) anticubes $\bigcup P_i^*$ with $\Gamma \subseteq \bigcup P_i^*$.

3.4.13 Notation

Let again $\mathcal{F}(\Gamma)$ be the set of all implicates of 1_Γ, further $\mathcal{F}_p(\Gamma) \subseteq \mathcal{F}(\Gamma)$ the set of all prime implicates of 1_Γ. Then we define

$$\mathcal{I}^*(\Gamma) := \{B \subseteq \mathcal{F}(\Gamma) : \min_{F \in B} F = 1_\Gamma\},$$

$$\mathcal{I}_p^*(\Gamma) := \{B \subseteq \mathcal{F}_p(\Gamma) : \min_{F \in B} F = 1_\Gamma\} \subseteq \mathcal{I}^*(\Gamma),$$

$$\mathcal{R}^*(\Gamma) := \{B \subseteq \mathcal{I}^*(\Gamma) : \min_{F \in K} F \neq 1_\Gamma \text{ for } K \subset B\} \subseteq \mathcal{I}^*(\Gamma),$$

$$\mathcal{R}_p^*(\Gamma) := \{B \subseteq \mathcal{I}_p^*(\Gamma) : \min_{F \in K} F \neq 1_\Gamma \text{ for } K \subset B\} = \mathcal{R}^*(\Gamma) \cap \mathcal{I}_p^*(\Gamma).$$

Now we state the implicate–counterpart to Theorem 2.4.15.

3.4.14 Theorem (see Theorem 2.4.15)

Let $F(\bar{a}_{(1)}),\dots,F(\bar{a}_{(m^*)})$ be the maximplicates of 1_Γ (corresponding to $\overline{\Gamma}=\{a_{(1)},\dots,a_{(m^*)}\}$). For each $\mu \in N_{m^*}$, define

$$\mathcal{M}^*_\mu(\Gamma) := \{F \in \mathcal{F}(\Gamma) : F \leq F(\bar{a}_{(\mu)})\},$$
$$\mathcal{M}^*_{p\mu}(\Gamma) := \{F \in \mathcal{F}_p(\Gamma) : F \leq F(\bar{a}_{(\mu)})\}$$

(thus

$$\mathcal{F}(\Gamma) = \bigcup_{N_{m^*}} \mathcal{M}^*_\mu(\Gamma), \mathcal{F}_p(\Gamma) = \bigcup_{N_{m^*}} \mathcal{M}^*_{p\mu}(\Gamma)).$$

Then

$$\mathcal{I}^*(\Gamma) = \mathcal{S}(\mathcal{M}^*_1(\Gamma),\dots,\mathcal{M}^*_{m^*}(\Gamma)),$$
$$\mathcal{I}^*_p(\Gamma) = \mathcal{S}(\mathcal{M}^*_{p1}(\Gamma),\dots,\mathcal{M}^*_{pm^*}(\Gamma)),$$
$$\mathcal{R}^*(\Gamma) = R(\mathcal{I}^*(\Gamma)),$$
$$\mathcal{R}^*_p(\Gamma) = R(\mathcal{I}^*_p(\Gamma)) = \mathcal{R}^*(\Gamma) \cap \mathcal{I}^*_p(\Gamma).$$

where $\mathcal{S}(A_1,\dots,A_{m^*})$ and $R(\mathcal{I})$ are defined by 2.4.13 and 2.4.14.

Chapter 4

Reduction Methods

In this chapter we state some rules which may be useful to determine the reduced set $R(\mathcal{S}(A_1, ..., A_m))$ for any finite sets $A_1, ..., A_m$ (see Definition 2.4.13, Definition 2.4.14 and Lemma 2.4.16). With $A_\mu = \mathcal{M}_\mu(\Gamma)$, $\mathcal{M}_{p\mu}(\Gamma), \mathcal{M}^*_\mu(\Gamma), \mathcal{M}^*_{p\mu}(\Gamma)$ respectively we may then determine the desired sets

$$\mathcal{R}(\Gamma), \mathcal{R}_p(\Gamma), \mathcal{R}^*(\Gamma), \mathcal{R}^*_p(\Gamma)$$

(see Theorem 2.4.15 and Theorem 3.4.14) with the help of this rules.

The last rule 4.8 is suitable to determine the minimal sets (sets with minimal size) of $R(\mathcal{S}(A_1, ..., A_m))$ corresponding to the minimal representations of 1_Γ by (prime) implicants respectively (prime) implicates.

We give the rules without proof.

4.1 Notation

For each K with $\emptyset \neq K = \{i_1, ..., i_k) \subseteq N_m$, we define R_K by $R_K := R(\mathcal{S}(A_{i_1}, ..., A_{i_k})) = R(\{B \subseteq \bigcup_K A_i : B \cap A_i \neq \emptyset \text{ for each } i \in K\})$, so that especially $R_{N_m} = R(\mathcal{S}(A_1, ..., A_m))$.

4.2 Rule

Let $K := \{i \in N_m : A_\mu \subset A_i \text{ for no } \mu \in N_m\}$. Then

$$R_{N_m} = R_K.$$

This rule obviously allows to omit all sets $A \in \{A_1, ..., A_m\}$ with $A_\mu \subset A$ for some $\mu \in N_m$. If $A_i \neq A_j$ for $i \neq j; i, j \in N_m$ then $\{A_j : j \in K\} = R(\{A_1, ..., A_m\})$.

Example

Assume $m = 4, A_1 = \{a_1, a_2\}, A_2 = \{a_2, a_3, a_4\}, A_3 = \{a_1, a_2, a_5\}$, $A_4 = \{a_2, a_3, a_4, a_6\}$. Then $K = \{1, 2\}$ and so $R_{\{1,2,3,4\}} = R_{\{1,2\}}$, i.e.

$$R(\mathcal{S}(A_1, ..., A_4)) = R(\mathcal{S}(A_1, A_2)) = R(\mathcal{S}(\{a_1, a_2\}, \{a_2, a_3, a_4\}))$$

(with $R(\mathcal{S}(\{a_1, a_2\}, \{a_2, a_3, a_4\})) = \{\{a_2\}, \{a_1, a_3\}, \{a_1, a_4\}\}$ as it is easy to see).

4.3 Rule

Define

$$K_1 := \{i \in N_m : |A_i| = 1\}, \quad K_2 := \{\mu \in N_m : A_\mu \cap \bigcup_{K_1} A_i = \emptyset\}.$$

Then, if $K_1 \neq \emptyset$,

$$R_{N_m} = \{A \cup \bigcup_{K_1} A_i : A \in R_{K_2}\}.$$

If $K_1 \neq \emptyset, K_2 = \emptyset$, then

$$R_{N_m} = \bigcup_{K_1} A_i.$$

Rule 4.3 says that in case $K_1 \neq \emptyset$ the union of all one–element–sets out of $\{A_1, ..., A_m\}$ appears in each element of R_{N_m}, moreover the problem to find R_{N_m} is reduced to the simpler problem to find R_{K_2}.

Example

Assume $m = 4, A_1 = \{a_1, a_2\}, A_2 = \{a_3\}, A_3 = \{a_4\}, A_4 = \{a_1, a_3, a_4\}$, $A_5 = \{a_2, a_5, a_6\}$. Then $K_1 = \{2,3\}, K_2 = \{1,5\}$ and so

$$R_{\{1,2,3,4,5\}} = \{A \cup \bigcup_{\{2,3\}} A_i : A \in R_{\{1,5\}}\} = \{A \cup \{a_3, a_4\} : A \in R_{\{1,5\}}\},$$

i.e.

$$R(S(A_1, ..., A_5)) = \{A \cup \{a_3, a_4\} : A \in R(S(\{a_1, a_2\}, \{a_2, a_5, a_6\}))\}$$

(with $R(S(\{a_1, a_2\}, \{a_2, a_5, a_6\})) = \{\{a_2\}, \{a_1, a_5\}, \{a_1, a_6\}\}$
and so

$$R(S(A_1, ..., A_4)) = \{\{a_2, a_3, a_4\}, \{a_1, a_3, a_4, a_5\}, \{a_1, a_3, a_4, a_6\}\}).$$

4.4 Rule

$$R_K = R(\{\{\omega_1\}, \cup \cdots \cup \{\omega_{|K|}\} : (\omega_1, ..., \omega_{|K|}) \in \mathop{\times}_{K} A_i\}).$$

This rule implies $|A| \leq |K|$ for each $A \in R_K$, especially $|A| \leq m$ for each $A \in R_{N_m}$.

4.5 Rule

Let $K \neq \emptyset, L \neq \emptyset, K \cap L = \emptyset$ and

$$R'_K := \{C \in R_K : D \subseteq C \text{ for some } D \in R_L\}$$
$$R'_L := \{C \in R_L : D \subseteq C \text{ for some } D \in R_K\}.$$

Then

$$R_{K \cup L} = R(R'_K \cup R'_L \cup \{A \cup B : (A, B) \in (R_K \setminus R'_K) \times (R_L \setminus R'_L)\}).$$

If $R'_K = R_K$ or $R_K \subseteq R_L$ then $R_{K \cup L} = R_K$,
if $R'_L = R_L$ or $R_L \subseteq R_K$ then $R_{K \cup L} = R_L$.
Clearly Rule 4.5 is only useful if R'_K or R'_L is not empty. In this case we obtain a simplification of the reduction procedure.

Note that $\subseteq$ may be replaced by $\subset$ in one of the sets R'_K, R'_L.

Example

$K = \{1,2,3\}, A_1 = \{a_3, a_4, a_5\}, A_2 = \{a_6, a_7\}, A_3 = \{a_1, a_3, a_4\}$
$L = \{4,5,6\}, A_4 = \{a_2, a_3\}, A_5 = \{a_1, a_5, a_7\}, A_6 = \{a_2, a_6\},$
$R_K = \{\{a_3, a_6\}, \{a_3, a_7\}, \{a_4, a_6\}, \{a_4, a_7\}, \{a_1, a_5, a_6\}, \{a_1, a_5, a_7\}\}$
$R_L = \{\{a_1, a_2\}, \{a_2, a_5\},, \{a_2, a_7\}, \{a_1, a_3, a_6\}, \{a_3, a_5, a_6\}, \{a_3, a_6, a_7\}\},$
$R'_K = \emptyset$
$R'_L = \{\{a_1, a_3, a_6\}, \{a_3, a_5, a_6\}, \{a_3, a_6, a_7\}\},$
$\{A \cup B : (A, B) \in (R_K \setminus R'_K) \times (R_L \setminus R'_L)\} =$
$\{\{a_2, a_3, a_7\}, \{a_2, a_4, a_7\}, \{a_1, a_2, a_3, a_6\}, \{a_1, a_2, a_4, a_6\}, \{a_1, a_2, a_5, a_6\},$
$\{a_1, a_2, a_5, a_7\}, \{a_2, a_3, a_5, a_6\}, \{a_2, a_4, a_5, a_6\}\},$
$R_{K \cup L} = \{\{a_1, a_3, a_6\}, \{a_3, a_5, a_6\}, \{a_3, a_6, a_7\}, \{a_2, a_3, a_7\}, \{a_2, a_4, a_7\}$
$\{a_1, a_2, a_4, a_6\}, \{a_1, a_2, a_5, a_6\}, \{a_1, a_2, a_5, a_7\}, \{a_2, a_4, a_5, a_6\}\}.$

4.6 Rule

For any $L \subseteq N_m (L \neq \emptyset)$, let $\{K_1, ..., K_r\}$ be a partition of L
(i.e. $\bigcup_{N_r} K_j = L, K_j \neq \emptyset, j = 1, ..., r; K_j \cap K_k = \emptyset$ for $j \neq k$). Then

$$R_L = R(\{\bigcup_{N_r} B_j : (B_1, ..., B_r) \in \bigtimes_{N_r} R_{K_j}\}).$$

The next rule is a consequence of Rule 4.6.

4.7 Rule (Corollary)

Let $\{K_1, ..., K_r\}$ be a partition of N_m (see Rule 4.6) and $L_\rho := \bigcup_{N_\rho} K_j, \rho = 1, ..., r$.

Then

$$R_{L_1} = R_{K_1};$$
$$R_{L_\rho} = R(\{A \cup B : (A, B) \in R_{L_{\rho-1}} \times R_{K_\rho}\}) \text{ for } \rho = 2, ..., r;$$
$$R_{L_r} = R_{N_m}.$$

With the aid of Rule 4.7 we may calculate R_{N_m} stepwise if $R_{K_1}, ..., R_{K_r}$ are already known.

Finally with the partition $\{\{1\}, ..., \{m\}\}$ of N_m from Rule 4.7 and Rule 4.5 we obtain the next result which allows to calculate R_{N_m} directly from the sets $A_1, ..., A_m$ by $m-1$ steps.

4.8 Rule

For $\mu = 1, ..., m-1$, define

$$R_{N_\mu}^{(0)} := \{A \in R_{N_\mu} : A \cap A_{\mu+1} = \emptyset\},$$
$$R_{N_\mu}^{(1)} := \{A \in R_{N_\mu} : A \cap A_{\mu+1} \neq \emptyset\}$$

(thus $R_{N_\mu}^{(0)} \cup R_{N_\mu}^{(1)} = R_{N_\mu}$, $R_{N_\mu}^{(0)} \cap R_{N_\mu}^{(1)} = \emptyset$).

Then for $\mu = 1, ..., m-1$

$$R_{N_{\mu+1}} = R_{N_\mu}^{(1)} \cup \{A \cup \{\omega\} : (A, \omega) \in R_{N_\nu}^{(0)} \times A_{\mu+1}, A' \subset A \cup \{\omega\} \text{ for no } A' \in R_{N_\mu}^{(1)}\}.$$

Remark

(a) $R_{N_1}^{(0)} = \{\{\omega\} : \omega \in A_1 \setminus A_2\}, R_{N_1}^{(1)} = \{\{\omega\} : \omega \in A_1 \cap A_2\}$.

(b) $\{A \cup \{\omega\} : (A, \omega) \in R_{N_\mu}^{(0)} \times A_{\mu+1},\ A' \subset A \cup \{\omega\}$ for no $A' \in R_{N_\mu}^{(1)}\}$
$= \{A \cup \{\omega\} : (A, \omega) \in R_{N_\mu}^{(0)} \times A_{\mu+1},\ A' \cap A_{\mu+1} = \{\omega\}$ and $A' \subset A \cup \{\omega\}$ for no $A' \in R_{N_\mu}^{(1)}\}$.

Remark (b) to Rule 4.8 suggests the following version of Rule 4.8.

4.8' Rule

Define $R^{(0)}_{N_\mu}$ and $R^{(1)}_{N_\mu}$ for $\mu = 1, ..., m-1$ as in Rule 4.8, further

$R^{(1,1)}_{N_\mu} := \{A \in R_{N_\mu} : |A \cap A_{\mu+1}| = 1\}$,

$A^*_{\mu+1} := \{\omega \in A_{\mu+1} : A \cap A_{\mu+1} \neq \{\omega\}$ for each $A \in R^{(1,1)}_{N_\mu}\}$,

$A^{**}_{\mu+1} := \{\omega \in A_{\mu+1} : A \cap A_{\mu+1} = \{\omega\}$ for some $A \in R^{(1,1)}_{N_\mu}\}$

(so that $R^{(1,1)}_{N_\mu} \subseteq R^{(1)}_{N_\mu}, A^*_{\mu+1} \cup A^{**}_{\mu+1} = A_{\mu+1}, A^*_{\mu+1} \cap A^{**}_{\mu+1} = \emptyset$).

Then

$$\begin{aligned} R_{N_{\mu+1}} = R^{(1)}_{N_\mu} &\cup \{A \cup \{\omega\} : (A,\omega) \in R^{(0)}_{N_\mu} \times A^*_{\mu+1}\} \\ &\cup \{A \cup \{\omega\} : (A,\omega) \in R^{(0)}_{N_\mu} \times A^{**}_{\mu+1}, A' \subset A \cup \{\omega\} \\ &\quad \text{for no } A' \in R^{(1,1)}_{N_\mu}\}. \end{aligned}$$

Sometimes we may be interested to find only reduced sets out of R_{N_m} with a given maximal size $\nu < m$ (if there are any) or even only such with minimal size (corresponding e.g. to minimal representations of binary functions by prime implicants or prime implicates).

The following last reduction rule now shows how to find for any given $\nu < m$ stepwise the sets $A \in R_{N_m}$ with $|A| \leq \nu$.

4.9 Rule

For $\nu \in N_{m-1}$ and $\mu \in N_m$, define

$R_{N_\mu,\nu} := \{A \in R_{N_\mu} : |A| \leq \nu\}, \mu = 1, ..., m.$

For $\mu \in N_{m-1}$, define

$R^{(0)}_{N_\mu,\nu} := \{A \in R_{N_\mu,\nu-1} : A \cap A_{\mu+1} = \emptyset\}$,

$R^{(1)}_{N_\mu,\nu} := \{A \in R_{N_\mu,\nu} : A \cap A_{\mu+1} \neq \emptyset\}$,

$R^{(1,1)}_{N_\mu,\nu} := \{A \in R_{N_\mu,\nu} : |A \cap A_{\mu+1}| = 1\}$,

$A^*_{\mu+1,\nu} := \{\omega \in A_{\mu+1} : A \cap A_{\mu+1} \neq \{\omega\}$ for each $A \in R^{(1,1)}_{N_\mu,\nu}\}$,

$A^{**}_{\mu+1,\nu} := \{\omega \in A_{\mu+1} : A \cap A_{\mu+1} = \{\omega\}$ for some $A \in R^{(1,1)}_{N_\mu,\nu}\}$

(so that $R^{(1,1)}_{N_\mu,\nu} \subseteq R^{(1)}_{N_\mu,\nu}, R^{(0)}_{N_\mu,\nu} \cap R^{(1)}_{N_\mu,\nu} = \emptyset, A^*_{\mu+1,\nu} \cup A^{**}_{\mu+1,\nu} = A_{\mu+1}$, $A^*_{\mu+1,\nu} \cap A^{**}_{\mu+1,\nu} = \emptyset$).

Then

$R_{N_\mu,\nu} = R_{N_\mu}$ for $\mu = 1, .., \nu$ (see Rules 4.8, 4.8') and

$$R_{N_{\mu+1},\nu} = R^{(1)}_{N_\mu,\nu} \cup \{A \cup \{\omega\} : (A,\omega) \in R^{(0)}_{N_\mu,\nu} \times A^*_{\mu+1,\nu}\}$$
$$\cup \{A \cup \{\omega\} : (A,\omega) \in R^{(0)}_{N_\mu,\nu} \times A^{**}_{\mu+1,\nu}, A' \subset A \cup \{\omega\}$$
$$\text{for no } A' \in R^{(1,1)}_{N_\mu,\nu}\}$$

for $\mu = \nu, ..., m-1$.

Rule 4.9 means that for each step from $R_{N_\mu,\nu}$ to $R_{N_{\mu+1},\nu}$ all $A \in R_{N_\mu,\nu}$ with $|A| = \nu$ and $A \cap A_{\mu+1} = \emptyset$ may be omitted.

We remark that $R_{N_\mu,\nu}$ may be empty for some μ. Clearly then $R_{N_{\mu+1},\nu}, ..., R_{N_m,\nu}$ are empty too. Thus Rule 4.9 is useful to prove whether there are (reduced) sets $B \in R_{N_\mu}$ with $|B| \leq \nu$.

If we are interested only in minimal sets out of R_{N_μ}, i.e. sets with minimal size, then we may begin to determine $R_{N_m,1}$. If $R_{N_m,1} \neq \emptyset$ then the problem has been solved. If not so we determine $R_{N_m,2}$ and so on until we find the first ν with $R_{N_m,\nu} \neq \emptyset$.

The next example is to show the efficiency of Rule 4.9 to detect minimal reduced sets.

Example

We assume (conf. Denis–Papin et al. (1974)) $A_1, ..., A_{18}$ to be given by

$$\begin{array}{lll} A_1 = \{6,12\} & A_7 = \{8,10\} & A_{13} = \{6,10,12\} \\ A_2 = \{1,7\} & A_8 = \{2,3\} & A_{14} = \{5,8,11\} \\ A_3 = \{1,5,9,11\} & A_9 = \{2,4\} & A_{15} = \{5,6,8,10\} \\ A_4 = \{1,5,6\} & A_{10} = \{2,12\} & A_{16} = \{1,7,9\} \\ A_5 = \{10,12\} & A_{11} = \{3,7\} & A_{17} = \{2,3,4\} \\ A_6 = \{8,11\} & A_{12} = \{4,9,11\} & A_{18} = \{3,4,7,9\} \end{array}$$

and so $A_1 \cup ... \cup A_{18} = \{1, ..., 12\}$.

Since

$$A_1 \subset A_{13}, A_6 \subset A_{14}, A_7 \subset A_{15}, A_2 \subset A_{16}, A_8 \subset A_{17}, A_{11} \subset A_{18},$$

we may omit $A_{13}, ..., A_{18}$ according to Rule 4.2. We like to determine $R_{N_{18},\nu} = R_{N_{12},\nu}$ with minimal ν.

Obviously the five sets $A_1, A_2, A_7, A_8, A_{12}$ are pairwise disjoint. Thus by definition $\mathcal{S}(A_1, ..., A_{12})$ contains no sets with less than five elements. Therefore $R_{N_{12},\nu} = \emptyset$

for $\nu = 1, ..., 4$. So we determine $R_{N_{12},5}$ using Rule 4.9 (resp. Rule 4.8 for $\nu = 1, ..., 5$).

Application of Rule 4.8 respectively 4.8' first yields

$$\begin{aligned}
R_{N_{1,5}} &= R_{N_1} = \{\{6\},\{12\}\},\\
R_{N_{2,5}} &= R_{N_2} = \{\{1,6\},\{1,12\},\{6,7\},\{7,12\}\},\\
R_{N_{3,5}} &= R_{N_3} = \{\{1,6\},\{1,12\},\{5,6,7\},\{6,7,9\},\{6,7,11\},\\
&\qquad \{5,7,12\},\{7,9,12\},\{7,11,12\}\}\\
R_{N_{4,5}} &= R_{N_4} = \{\{1,6\},\{1,12\},\{5,6,7\},\{6,7,9\},\{6,7,11\},\{5,7,12\}\},\\
R_{N_{5,5}} &= R_{N_5} = \{\{1,12\},\{1,6,10\},\{5,7,12\},\{5,6,7,10\},\\
&\qquad \{6,7,9,10\},\{6,7,9,12\},\{6,7,10,11\},\{6,7,11,12\}\}.
\end{aligned}$$

Now for $\mu = 4, ..., 12$ we may apply Rule 4.9.

$$\begin{aligned}
R^{(0)}_{N_5,5} &= \{\{1,12\},\{1,6,10\},\{5,7,12\},\{5,6,7,10\},\{6,7,9,10\},\{6.7.9,12\}\},\\
R^{(1)}_{N_5,5} &= R^{(1,1)}_{N_5,5} = \{\{6,7,10,11\},\{6,7,11,12\}\\
A^{*}_{6,5} &= \{8\}, A^{**}_{6,5} = \{11\}\\
\Rightarrow R_{N_6,5} &= \{6,7,10,11\},\{6,7,11,12\}\}\\
&\quad \cup\{\{1,8,12\},\{1,6,8,10\},\{5,7,8,12\},\{5,6,7,8,10\},\{6,7,8,9,10\},\\
&\qquad \{6,7,8,9,12\}\}\\
&\quad \cup\{\{1,11,12\},\{1,6,10,11\},\{5,7,11,12\}\},\\
R^{(0)}_{N_6,5} &= \{\{1,11,12\},\{5,7,11,12\},\{6,7,11,12\}\},\\
R^{(1)}_{N_6,5} &= \{\{1,8,12\},\{1,6,8,10\},\{1,6,10,11\},\{5,7,8,12\},\{6,7,10,11\},\\
&\qquad \{5,6,7,8,10\},\{6,7,8,9,10\},\{6,7,8,9,12\}\},\\
R^{(1,1)}_{N_6,5} &= \{\{1,8,12\},\{1,6,10,11\},\{5,7,8,12\},\{6,7,10,11\},\{6,7,8,9,12\}\}.\\
A^{*}_{7,5} &= \emptyset, A^{**}_{7,5} = \{8,10\}\\
\Rightarrow R_{N_7,5} &= \{\{1,8,12\},\{1,6,7,10\},\{1,6,10,11\},\{5,7,8,12\},\{6,7,10,11\},\\
&\qquad \{5,6,7,8,10\},\{6,7,8,9,10\},\{6,7,8,9,12\}\}\\
&\quad \cup\{\{1,10,11,12\},\{5,7,10,11,12\},\{6,7,8,11,12\}\},\\
R^{(0)}_{N_7,5} &= \{\{1,8,12\},\{1,6,8,10\},\{1,6,10,11\},\{5,7,8,12\},\{6,7,10,11\},\\
&\qquad \{1,10,11,12\}\},\\
R^{(1)}_{N_7,5} &= R^{(1,1)}_{N_7,5} = \emptyset,\\
A^{*}_{8,5} &= A_8 = \{2,3\}, A^{**}_{8,5} = \emptyset\\
\Rightarrow R_{N_8,5} &= \{\{1,2,8,12\},\{1,2,6,8,10\},\{1,2,6,10,11\},\{2,5,7,8,12\}
\end{aligned}$$

$$\{2,6,7,10,11\},\{1,2,10,11,12\},\{1,3,8,12\},\{1,3,6,8,10\},$$
$$\{1,3,6,10,11\},\{3,5,7,8,12\},\{3,6,7,10,11\},\{1,3,10,11,12\}\},$$
$$R^{(0)}_{N_8,5} = \{\{1,3,8,12\}\},$$
$$R^{(1)}_{N_8,5} = R^{(1,1)}_{N_8,5} = \{\{1,2,8,12\},\{1,2,6,8,10\},\{1,2,6,10,11\},\{2,5,7,8,12\},$$
$$\{2,6,7,10,11\},\{1,2,10,11,12\}\},$$
$$A^*_{9,5} = \{4\}, A^{**}_{9,5} = \{2\}$$
$$\Rightarrow R_{N_9,5} = \{\{1,2,8,12\},\{1,2,6,8,10\},\{1,2,6,10,11\},\{2,5,7,8,12\},$$
$$\{2,6,7,10,11\},\{1,2,10,11,12\}\}$$
$$\cup\{\{1,3,4,8,12\}\},$$
$$R^{(0)}_{N_9,5} = \emptyset, R^{(1)}_{N_9,5} = R_{N_9,5}$$
$$\Rightarrow R_{N_{10},5} = R_{N_9,5},$$
$$R^{(0)}_{N_{10},5} = \{\{1,2,8,12\}\},$$
$$R^{(1)}_{N_{10},5} = R^{(1,1)}_{N_{10},5} = \{\{2,5,7,8,12\},\{2,6,7,10,11\},\{1,3,4,8,12\}\}$$
$$A^*_{11,5} = \{3\}, A^{**}_{11,5} = \{7\}$$
$$\Rightarrow R_{N_{11},5} = \{\{2,5,7,8,12\},\{2,6,7,10,11\},\{1,3,4,8,12\}\}$$
$$\cup\{\{1,2,3,8,12\}\}\cup\{\{1,2,7,8,12\}\},$$
$$R^{(0)}_{N_{11},5} = \emptyset, R^{(1)}_{N_{11},5} = \{\{1,3,4,8,12\},\{2,6,7,10,11\}\}$$
$$\Rightarrow R_{N_{12},5} = \{\{1,3,4,8,12\},\{2,6,7,10,11\}\}.$$

Thus we have two minimal elements of

$$R(S(A_1,...,A_{18})) = R(S(A_1,...,A_{12})),$$

namely $\{1,3,4,8,12\}$ and $\{2,6,7,10,11\}$.

Chapter 5

Discrete Functions

In this chapter we consider *discrete functions*

(a) $f : M \to [0,\infty), f(M) = \{0, y_1, ..., y_k\}$ with $0 < y_1 < ... < y_k, k \in \mathbb{N}$.

Thus a discrete function is a function of n finite–valued variables and takes the $k+1$ different nonnegative values $0, y_1, ..., y_k$. In case $f(M) = \{0,1\}$ especially f is a binary function.

Obviously every function $g : M \to \mathbb{R}$ with $g(M) = \{z_0, z_1, ..., z_k\}$, $z_0 < ... < z_k$ may be written as

$$g = z_0 + f$$

where f is a discrete function with $f(M) = \{0, z_1 - z_0, ..., z_k - z_0\}$. Thus (a) means no essential loss of generality.

In the following f denotes a discrete function according to (a).

5.1 Representations by Binary Functions

First we show that f has several representations by the indicators $1_{\{f=y_i\}}, 1_{\{f\geq y_i\}}$, $1_{\{f\leq y_i\}}, i \in \{1, ..., k\}$ defined by

$$1_{\{f=y_i\}}(x) := \begin{cases} 1 & \text{for } f(x) = y_i \\ 0 & \text{otherwise,} \end{cases}$$

$$1_{\{f\geq y_i\}}(x) := \begin{cases} 1 & \text{for } f(x) \geq y_i \\ 0 & \text{otherwise,} \end{cases}$$

$$1_{\{f\leq y_i\}}(x) := \begin{cases} 1 & \text{for } f(x) \leq y_i \\ 0 & \text{otherwise.} \end{cases}$$

5.1.1 Theorem

The discrete function f has the representations

(5.1.2) $f = \sum_{i=1}^{k} y_i 1_{\{f=y_i\}}$,

(5.1.3) $f = \max_{1\leq i\leq k} (y_i 1_{\{f=y_i\}})$,

(5.1.4) $f = \sum_{i=1}^{k} (y_i - y_{i-1}) 1_{\{f\geq y_i\}}$
$= \sum_{i=0}^{k-1} (y_{i+1} - y_i) 1_{\{f>y_i\}}$ with $y_0 := 0$,

(5.1.5) $f = \max_{1\le i\le k}(y_i 1_{\{f\ge y_i\}})$,

(5.1.6) $f = \min_{1\le i\le k}(y_{i-1} + (y_k - y_{i-1})1_{\{f\ge y_i\}})$

$= \min_{0\le i\le k-1}(y_i + (y_k - y_i))1_{\{f> y_i\}}$.

Proof

For $x \in M$ with $f(x) = 0$, we have

$$1_{\{f=y_i\}}(x) = 1_{\{f\ge y_i\}}(x) = 0 \text{ for } i = 1, \ldots, k,$$

thus all terms on the right hand side are equal 0.

For $x \in M$ with $f(x) = y_j, j \in \{1, \ldots, k\}$, we have

$$1_{\{f=y_i\}}(x) = \begin{cases} 1 & \text{for } i = j \\ 0 & \text{otherwise} \end{cases}$$

and

$$1_{\{f\ge y_i\}}(x) = \begin{cases} 1 & \text{for } 1 \le i \le j \\ 0 & \text{otherwise} \end{cases}$$

implying

$$\sum_{i=1}^{k} y_i 1_{\{f=y_i\}}(x) = \max_{1\le i\le k}(y_i 1_{\{f=y_i\}}(x)) = y_j = f(x),$$

$$\sum_{i=1}^{k}(y_i - y_{i-1})1_{\{f\ge y_i\}}(x) = \sum_{i=1}^{j}(y_i - y_{i-1}) = y_j = f(x),$$

$$\max_{1\le i\le k}(y_i 1_{\{f\ge y_i\}}(x)) = \max_{1\le i\le j} y_i = y_j = f(x),$$

further, if $j < k$,

$$y_{i-1} + (y_k - y_{i-1})1_{\{f\ge y_i\}}(x) = \begin{cases} y_k & \text{for } 1 \le i \le j \\ y_{i-1} & \text{for } i > j \end{cases}$$

and so

$$\min_{1\le i\le k}(y_{i-1} + (y_k - y_{i-1})1_{\{f\ge y_i\}}(x)) = \min(y_k, y_j) = y_j = f(x).$$

Finally for $j = k$ (i.e. $f(x) = y_k$), we have

$$y_{i-1} + (y_k - y_{i-1})1_{\{f\ge y_i\}}(x) = y_k \text{ for } 1 \le i \le k$$

and so

$$\min_{1\le i\le k}(y_{i-1} + (y_k - y_{i-1})1_{\{f\ge y_i\}}(x)) = y_k = f(x). \qquad \square$$

Corollary

If $y_i = i$ for $i \in \{1, ..., k\}$ then

(5.1.7) $f = \sum_{i=1}^{k} i 1_{\{f=i\}}$,

(5.1.8) $f = \max_{1 \le i \le k} (i 1_{\{f=i\}})$,

(5.1.9) $f = \sum_{i=1}^{k} 1_{\{f \ge i\}} = \sum_{i=0}^{k-1} 1_{\{f > i\}}$

(5.1.10) $f = \max_{1 \le i \le k} (i 1_{\{f \ge i\}})$,

(5.1.11) $f = \min_{1 \le i \le k} (i - 1 + (k - i + 1) 1_{\{f \ge i\}}) = \min_{0 \le i \le k-1} (i + (k - i) 1_{\{f > i\}})$.

Remark

Using $1_{\{f \ge y_i\}} = 1 - 1_{\{f < y_i\}} = 1 - 1_{\{f \le y_{i-1}\}}$, we obtain from (5.1.4) and (5.1.6)

(5.1.12) $f = y_k - \sum_{i=0}^{k-1} (y_{i+1} - y_i) 1_{\{f \le y_i\}}$,

(5.1.13) $f = \min_{0 \le i \le k-1} (y_k - (y_k - y_i) 1_{\{f \le y_i\}})$.

By Theorem 5.1.1 we obtained representations of a discrete function f by the binary functions (indicators) $1_{\{f=y_i\}}$ and $1_{\{f \ge y_i\}}, i \in \{1, ..., k\}$.

Now we may represent these indicators by their implicants (prime implicants) or by their implicates (prime implicates). So we obtain f as an expression in such special indicator implicants or implicates.

Consider for instance minimal representations

$$1_{\{f=y_i\}} = \max(C_{i1}, ..., C_{i\mu_i}), i = 1, ..., k$$

by prime implicants. Then (5.1.2) implies

$$f = \sum_{i=1}^{k} y_i \max(C_{i1}, ..., C_{i\mu_i}) = \max_{i \in N_k} (y_i \max(C_{i1}, ..., C_{i\mu_i}))$$
$$= \max_{\substack{i \in N_k \\ j \in N_{\mu_i}}} (y_i C_{ij})$$

where the functions $y_i C_{ij}$ satisfy the condition

$$y_i C_{ij} \le f.$$

Thus we may regard them as special "implicants" of f, because C_{ij} is an implicant of $1_{\{f=y_i\}}$ and so $C_{ij} = 1$ implies $C_{i'j} = 0$ for all $C_{i'j}$ with $i' \ne i$. In fact sometimes such a representation may be useful, first of all for the class of monotone functions and for the more general class of semimonotone functions, both treated in the next sections.

5.2 Monotone Functions

For monotone discrete functions the indicators $1_{\{f=y\}}, 1_{\{f\geq y\}}, 1_{\{f\leq y\}}$ have special representations as maxima of prime implicants which are interval indicators. Moreover monotone functions have special unique representations by prime implicants also related to interval indicators. Finally there are corresponding representations by implicates related to complements of intervals. Thus we first treat monotone functions .

5.2.1 Definition

1. A discrete function is called *isotone* if $x \leq x'$ (i.e. $x_i \leq x_i'$ for $i = 1, ..., n$) implies $f(x) \leq f(x')$.
2. A discrete function is called *antitone* if $x \leq x'$ implies $f(x) \geq f(x')$.
3. A discrete function is called *monotone* if it is either isotone or antitone.
4. Let $x, x' \in M, x \leq x'$. Then we call

$$[x_i, x_i'] := \{z_i \in M_i : x_i \leq z_i \leq x_i'\} \subseteq M_i$$

an *interval of* M_i, $i \in N_n$, and

$$[x, x'] := \times [x_i, x_i'] \subseteq M$$

an *interval of* M.

5.2.2 Lemma

Let $P_i, Q_i \subseteq M_i, i \in N$. Then the following are equivalent:

(a) $\times P_i \subset \times Q_i$.
(b) $P_i \subseteq Q_i$ for $i \in N_n$, $P_i \subset Q_i$ for at least one $i \in N_n$.

Proof

The assertion is evident.

5.2.3 Corollary

Let $[x, x']$ and $[z, z']$ be intervals of M. Then the following are equivalent

(a) $[x, x'] \subset [z, z']$.
(b) It holds $z < x$ and $x' < z'$
 or $z < x$ and $x' = z'$
 or $z = x$ and $x' < z$.

Next we need the idea of maximal points and minimal points of a subset Q of M.

5.2.4 Definition

For any $Q \subseteq M$, the point $x \in Q$ is called a *minimal point* of Q if there is no $x' \in Q$ with $x' < x$, and $x \in Q$ is called a *maximal point* of Q if there is no $x' \in Q$ with $x' > x$.
The minimal point of M is $0 := (0, ..., 0)$, the maximal point of M is $a^* := (a_{1k_1}, ..., a_{n,k_n})$ (see p. 3).

5.2.5 Theorem

Let f be isotone, $y \in \{y_1, ..., y_k\}$ and $G_{\geq y}$ the set of all minimal points of $\{f \geq y\}$. Then the set $C_p(\{f \geq y\})$ of all prime implicants of $1_{\{f \geq y\}}$ is given by

(5.2.6) $$C_p(\{f \geq y\}) = \{1_{[z,a^*]} : z \in G_{\geq y}\}.$$

The only representation of $1_{\{f \geq y\}}$ by prime implicants is given by the maximum of all its prime implicants:

(5.2.7) $$1_{\{f \geq y\}} = \max_{C \in C_p(\{f \geq y\})} C = \max_{z \in G_{\geq y}} 1_{[z,a*]}.$$

This means that in (5.2.7) no $C \in C_p(\{f \geq y\})$ may be omitted.

Proof

a) Suppose $C = 1_{[z,a*]}$ with $z \in G_{\geq y}$. We show that C is a prime implicant of $1_{\{f \geq y\}}$. The isotony of f implies that C is an implicant of $1_{\{f \geq y\}}$. Assume C' to be a prime implicant of $1_{\{f \geq y\}}$ with $C < C'$. Then C' is an interval indicator due to the isotony of f, i.e. $C' = 1_{[z',a*]}$ with $z' < z$ according to Corollary 5.2.3. But then $z' \notin 1_{\{f \geq y\}}$ since z is a minimal point of $\{f \geq y\}$. Thus C' cannot be an implicant of $1_{\{f \geq y\}}$ and so C is a prime implicant of $1_{\{f \geq y\}}$.
b) Suppose C to be a prime implicant of $1_{\{f \geq y\}}$. Then $C = 1_{[z,z^*]}$. Assume $z \notin G_{\geq y}$ or $z^* < a^*$. Then $C' = 1_{[z',a^*]}$ with $z' \in G_{\geq y}, z' \leq z$ is an implicant of $1_{\{f \geq y\}}$ with $C < C'$ in contradiction to the supposition. Thus $z \in G_{\geq y}$ and $z' = a^*$.

c) We have still to proove that in (5.2.7) no $C \in \mathcal{C}_p(\{f \geq y\})$ may be omitted. If $\mathcal{C}_p(\{f \geq y\})$ contains only one prime implicant i.e. if $1_{\{f \geq y\}}$ is an interval indicator itself, then evidently the proposition holds. Now we suppose $|G_{\geq y}| > 1$ and $z' \in G_{\geq y}$. Then for each $z \in G_{\geq y}, z \neq z'$ it holds $z' \notin [z, a^*]$ and so $1_{[z,a^*]}(z') = 0$, thus

$$\max_{\substack{C \in \mathcal{C}_p(\{f \geq y\}) \\ C \neq 1_{[z',a^*]}}} C(z') = 0 < 1_{\{f \geq y\}}(z') = 1,$$

and so $1_{[z',a^*]}$ cannot be omitted in (5.2.7). □

If f is an antitone discrete function we may state the following counterpart of Theorem 5.2.5.

5.2.8 Theorem

Let f be antitone, $y \in \{y_1, ..., y_k\}$ and $G^*_{\geq y}$ the set of all maximal points of $\{f \geq y\}$. Then the set $\mathcal{C}_p(\{f \geq y\})$ of all prime implicants of $1_{\{f \geq y\}}$ is given by

$$\text{(5.2.9)} \qquad \mathcal{C}_p(\{f \geq y\}) = \{1_{[0,z]} : z \in G^*_{\geq y}\}.$$

The only representation of $1_{\{f \geq y\}}$ by prime implicants is given by the maximum of all its prime implicants:

$$\text{(5.2.10)} \qquad 1_{\{f \geq y\}} = \max_{C \in \mathcal{C}_p(\{f \geq y\})} C = \max_{z \in G^*_{\geq y}} 1_{[0,z]}.$$

Proof

The proof is similar to that of Theorem 5.2.5.

The next theorems give the prime implicates of the indicators $1_{\{f > y\}}$ and the representations of $1_{\{f > y\}}$ by prime implicates.

5.2.11 Theorem

Let f be isotone, $y \in \{0, ..., y_{k-1}\}, G^*_{\leq y}$ the set of all maximal points of $\{f \leq y\}$. Then the set $\mathcal{F}_p(\{f > y\})$ of all prime implicates of $1_{\{f > y\}}$ is given by

$$\text{(5.2.12)} \qquad \mathcal{F}_p(\{f > y\}) = \{1_{\overline{[0,z]}} : z \in G^*_{\leq y}\} = \{1 - 1_{[0,z]} : z \in G^*_{\leq y}\}.$$

The only representation of $1_{\{f>y\}}$ by prime implicates is given by the minimum (product) of all its prime implicates:

$$(5.2.13)\qquad 1_{\{f>y\}} = \min_{f\in\mathcal{F}_p(\{f>y\})} F = \min_{z\in G^*_{\leq y}} 1_{\overline{[0,z]}} = \prod_{z\in G^*_{\leq y}} 1_{\overline{[0,z]}}\,.$$

Proof

It is easy to see that $\{1_{[0,z]} : z \in G^*_{\leq y}\}$ is the set of all prime implicants of $1_{\{f\leq y\}}$, further that

$$(5.2.14)\qquad 1_{\{f\leq y\}} = \max_{z\in G^*_{\leq y}} 1_{[0,z]}$$

is the only representation of $1_{\{f\leq y\}}$ by prime implicants. (The proof is similar to that of Theorem 5.2.5). From Theorem 3.3.2 and (3.1.2) we see that any anticube indicator F is a prime implicate of $1_{\{f>y\}}$ if and only if $1 - F$ is prime implicant of $1_{\{f\leq y\}}$. Thus (5.2.12) holds. The unique representation (5.2.13) follows from

$$\begin{aligned} 1_{\{f>y\}} &= 1 - 1_{\{f\leq y\}} = 1 - \max_{z\in G^*_{\leq y}} 1_{[0,z]} = \min_{z\in G^*_{\leq y}} (1 - 1_{[0,z]}) \\ &= \min_{z\in G_{\leq y}} 1_{\overline{[0,z]}} \end{aligned}$$

and the uniqeness of (5.2.14). □

In the same way we may prove the next result.

5.2.15 Theorem

Let f be antitone, $y \in \{0, ..., y_{k-1}\}$ and $G_{\leq y}$ the set of all minimal points of $\{f \leq y\}$. Then the set $\mathcal{F}_p(\{f > y\})$ of all prime implicates of $1_{\{f>y\}}$ is given by

$$(5.2.16)\qquad \begin{aligned} \mathcal{F}_p(\{f > y\}) &= \{1_{\overline{[z,a^*]}} : z \in G_{\leq y}\} \\ &= \{1 - 1_{[z,a^*]} : z \in G_{\leq y}\}. \end{aligned}$$

The only representation of $1_{\{f>y\}}$ by prime implicates is given by the minimum (product) of *all* prime implicates:

$$1_{\{f>y\}} = \min_{F\in\mathcal{F}_p(\{f>y\})} F = \min_{z\in G_{\leq y}} 1_{\overline{[z,a^*]}} = \prod_{z\in G_{\leq y}} 1_{\overline{[z,a^*]}}.$$

By the last theorems we obtained unique representations of the indicators $1_{\{f\geq y\}}$, $1_{\{f>y\}}$ by prime implicants respectively prime implicates, where f was assumed to be monotone. Now we give unique representations of a monotone discrete function f itself.

5.2.17 Theorem

a) Let f be isotone, G_{y^*} the set of all minimal points of $\{f = y\}, y \in \{y_1, ..., y_k\}$. Then f has the unique representation

$$f = \max_{\substack{i \in N_k \\ z \in G_{y_i^*}}} y_i 1_{[z,a^*]} \tag{5.2.18}$$

(so that no term $y_i 1_{[z,a^*]}$ in (5.2.18) may be omitted).

b) Let f be antitone, G_y^* the set of all maximal points of $\{f = y\}$, $y \in \{y_1, ..., y_k\}$. Then f has the unique representation

$$f = \max_{\substack{i \in N_k \\ z \in G_{y_i}^*}} y_i 1_{[0,z]}. \tag{5.2.19}$$

Proof

a) For $x \in \{f = 0\}$ we have $1_{[z,a^*]}(x) = 0$ for all $z \in G_{y_i^*}$ with $i \in N_k$ and so

$$\max_{\substack{i \in N_k \\ z \in G_{y_i^*}}} y_i 1_{[z,a^*]}(x) = 0 = f(x).$$

For $x \in \{f = y_j\}, j \in N_k$ there is an $z \in G_{y_j^*}$ with $x \in [z, a^*]$ and thus $y_j 1_{[z,a^*]}(x) = y_j = f(x)$. Further

$$y_i 1_{[z,a^*]}(x) \begin{cases} \leq y_j & \text{for } i \leq j, \\ 0 & \text{for } i > j, z \in G_{y_i^*} \end{cases}$$

and so

$$\max_{\substack{i \in N_k \\ z \in G_{y_i^*}}} y_i 1_{[z,a^*]}(x) = y_j = f(x).$$

Thus (5.2.18) holds.

To prove the uniqueness we choose any $y_j 1_{[z',a^*]}$ with $j \in N_k, z' \in G_{y_j^*}$ and show that $y_j 1_{[z',a^*]}$ cannot be omitted in (5.2.18). From $z' \in G_{y_j^*}$ it follows $f(z') = y_j$, further

$$y_i 1_{[z,a^*]}(z') \begin{cases} < y_j & \text{for } i < j \\ = 0 & \text{for } i = j, z \in G_{y_j^*}, z \neq z' \\ = 0 & \text{for } i > j, z \in G_{y_i^*} \end{cases}$$

implying

$$\max_{\substack{i \in N_k \\ z \in G_{y_i^*} \\ z \neq z'}} y_i 1_{[z,a^*]}(z') \neq y_j = f(z').$$

Thus $y_j 1_{[z',a^*]}$ cannot be omitted in (5.2.18).

b) The proof of b) is the same. □

Remark

In Theorem 4.4.5 we will see that Theorem 5.2.17 states the unique representation of a monotone discrete function as the maximum of all its prime implicants (see Definition in Section 5.4).

5.2.20 Theorem

a) Let f be isotone. Then f has the unique representation

(5.2.21) $$f = \min_{\substack{0 \leq i \leq k-1 \\ z \in G^*_{y_i}}} (y_i + (y_k - y_i) 1_{\overline{[0,z]}}).$$

b) Let f be antitone. Then f has the unique representation

(5.2.22) $$f = \min_{\substack{0 \leq i \leq k-1 \\ z \in \overline{G}_{y_i^*}}} (y_i + (y_k - y_i) 1_{\overline{[z,a^*]}}).$$

Proof

The proof is similar to that of Theorem 5.2.17.

Remark

In Theorem 5.6.12 we will see that Theorem 5.2.20 states the unique representation of a monotone discrete function as the minimum of all its prime implicates (see Definition in Section 5.6).

5.3. Semimonotone Functions

5.3.1 Definition

A discrete function is called *semimonotone* if $x \leq x'$, $f(x) = f(x')$ implies $f(z) = f(x)$ for all $z \in [x, x']$.

We remark that every monotone function is semimonotone too.

Now for any semimonotone function f and any $y \in f(M)$ we state the set of all prime implicants of $1_{\{f=y\}}$.

5.3.2 Theorem

Let f be semimonotone and $y \in f(M)$. Let G_{y*} be the set of all minimal points and G_y^* the set of all maximal points of $\{f = y\}$. Then the set $\mathcal{C}_p(\{f = y\})$ of all prime implicants of $1_{\{f=y\}}$ is given by

$$\mathcal{C}_p(\{f = y\}) = \{1_{[z,z']} : (z, z') \in G_{y*} \times G_y^*\}. \tag{5.3.3}$$

Thus a cube indicator is a prime implicant of $1_{\{f=y\}}$ if and only if the cube is an interval $[z, z']$ with a minimal point z of $\{f = y\}$ and a maximal point z' of $\{f = y\}$.

Proof

a) Suppose $C = 1_{[z,z']}$ with $(z, z') \in G_{y*} \times G_y^*$. We show that C is a prime implicant of $1_{\{f=y\}}$. The semimonotony of f implies that C is a prime implicant of $1_{\{f=y\}}$ with $C < C'$. Then again C' has the form $C' = 1_{[t,t']}$ with $t, t' \in \{f = y\}$ due to the semimonotony of f. From $C < C'$ it follows $[z, z'] \subset [t, t']$ and so

$$\begin{aligned} &t < z \text{ and } z' < t' \\ \text{or } &t < z \text{ and } z' = t' \\ \text{or } &t = z \text{ and } z' < t' \end{aligned}$$

due to Corollary 5.2.3. This is impossible since z is a minimal point and z' a maximal point of $\{f = y\}$. Thus there is no prime implicant C' of $1_{\{f=y\}}$ with $C < C'$ and thus C is a prime implicant of $1_{\{f=y\}}$.

b) Suppose C to be a prime implicant of $1_{\{f=y\}}$ and so $C = 1_{[z,z']}$ with $z, z' \in \{f = y\}$. Assume $z \notin G_{y*}$. Then there is a $t \in G_{y*}$ with $t < z$ and so $C' = 1_{[t,z']}$ is an implicant of $1_{\{f=y\}}$ with $C < C'$ in contradiction to the suppostion. Thus $z \in G_{y*}$. In the same way it follows $z' \in G_y^*$. □

Theorem 5.3.2 immediately yields the set $\mathcal{C}_p(\{f = y\})$ of all prime implicants of $1_{\{f=y\}}$. All these prime implicants are interval indicators. We now may use the methods of Section 2.4 to give reduced (minimal) representations of the indicators $1_{\{f=y_i\}}, i \in N_n$ by prime implicants. Finally from (5.1.2) or (5.1.3) we obtain a sum representation or a maximum representation of a semimonotone discrete function f with the help of the prime implicants of the several indicators $1_{\{f=y_i\}}, i \in N_n$. We remark that in general the indicators $1_{\{f=y\}}$ have no unique representations by prime implicants.

Example

Let $n = 2, M_1 = M_2 = \{0,1,2,3,4\} = [0,4], M = [0,4] \times [0,4]$ and the (isotone) discrete function $f : M \to \{0,1,2,3,4\}$ defined by

$$f(x_1,x_2) := \begin{cases} 0 & \text{for } (x_1,x_2) \in \{(0,0),(0,1),(0,2),(1,0)\}, \\ 1 & \text{for } (x_1,x_2) \in \{(0,3),(0,4),(1,1),(1,2),(2,0),(2,1),(3,0)\}, \\ 2 & \text{for } (x_1,x_2) \in \{(1,3),(1,4),(2,2),(2,3),(2,4),(3,1),(3,2), \\ & \qquad (4,0),(4,1)\} \\ 3 & \text{for } (x_1,x_2) \in \{(3,3),(4,2)\}, \\ 4 & \text{for } (x_1,x_2) \in \{(3,4),(4,3),(4,4)\} \end{cases}$$

corresponding to the following scheme:

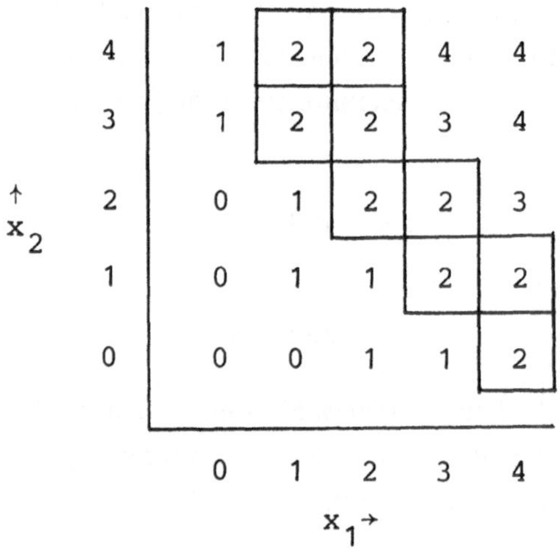

We consider the indicator $1_{\{f=2\}}(x_1,x_2)$ taking the value 1 on the 9 points $(1,3),(1,4),(2,2),(2,3),(2,4),(3,1),(3,2),(4,0),(4,1)$ and the value 0 otherwise on $M = [0,4] \times [0,4]$.

Minimal points of $1_{\{f=2\}}$ are $(1,3),(2,2),(3,1),(4,0)$.

Maximal points of $1_{\{f=2\}}$ are $(2,4),(3,2),(4,1)$.

Prime implicants of $1_{\{f=2\}}$ are $1_{I_1}, 1_{I_2}, 1_{I_3}, 1_{I_4}, 1_{I_5}, 1_{I_6}$ with

$I_1 = [1,2] \times [3,4], I_2 = [2] \times [2,4]$,

$I_3 = [2,3] \times [2] \quad , I_4 = [3] \times [1,2]$,

$I_5 = [3,4] \times [1] \quad , I_6 = [4] \times [0,1]$.
Representations of $1_{\{f=2\}}$ by prime implicants are

$$\begin{aligned} 1_{\{f=2\}} &= \max(1_{I_1}, 1_{I_2}, 1_{I_3}, 1_{I_4}, 1_{I_5}, 1_{I_6}) \\ &= \max(1_{I_1}, 1_{I_2}, 1_{I_3}, 1_{I_4}, 1_{I_6}) \\ &= \max(1_{I_1}, 1_{I_2}, 1_{I_3}, 1_{I_5}, 1_{I_6}) \\ &= \max(1_{I_1}, 1_{I_2}, 1_{I_4}, 1_{I_5}, 1_{I_6}) \\ &= \max(1_{I_1}, 1_{I_3}, 1_{I_4}, 1_{I_5}, 1_{I_6}) \\ &= \max(1_{I_1}, 1_{I_3}, 1_{I_4}, 1_{I_6}). \end{aligned}$$

The reduced representations are

$$\begin{aligned} 1_{\{f=2\}} &= \max(1_{I_1}, 1_{I_2}, 1_{I_3}, 1_{I_5}, 1_{I_6}) \\ &= \max(1_{I_1}, 1_{I_2}, 1_{I_4}, 1_{I_5}, 1_{I_6}) \\ &= \max(1_{I_1}, 1_{I_3}, 1_{I_4}, 1_{I_6}). \end{aligned}$$

The unique minimal representation is

$$1_{\{f=2\}} = \max(1_{I_1}; 1_{I_3}, 1_{I_4}, 1_{I_6}).$$

The illustration of this minimal representation of $1_{\{f=2\}}$ as a parallel circuit of four series circuits according to (2.4.3) is given by the following figure:

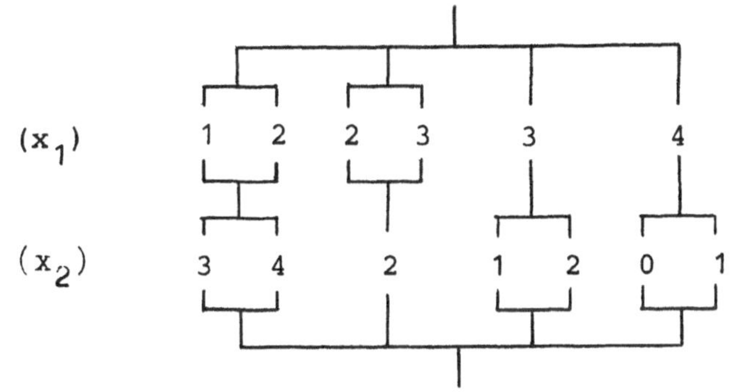

From section 3.4 we know that $1_{\{f=2\}}$ has representations by *prime implicates* too. These representations have illustrations as series circuits of parallel circuits according to (3.4.3). We give such a circuit by the next figure. The proof is left to the reader.

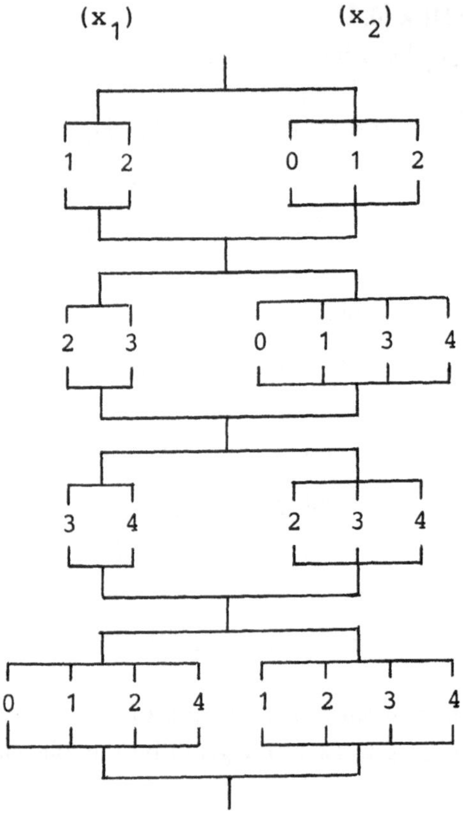

5.4 Implicants (Prime Implicants) of Discrete Functions

The unique representations of monotone discrete functions given by Theorem 5.2.17 and Theorem 5.2.20 suggest to introduce implicants and implicates of general discrete functions (see Davio et al. (1978)).

Thus in this section we define implicants (prime implicants) of a discrete function. Then we will show that we obtain these from implicants of related indicator implicants. Finally we will state representations of discrete functions as maxima of implicants (prime implicants).

5.4.1 Definition

Let f be a discrete function, C a cube indicator on M (Definition 2.1.1) and $y \in \{y_1, ..., y_k\}$. Then

$$I(y, C) := \min(y, yC) = yC = \begin{cases} y & \text{for } C = 1 \\ 0 & \text{for } C = 0 \end{cases}$$

is called an *implicant* of f if $yC \leq f$ and $yC(x_0) = f(x_0) = y$ for at least one $x_0 \in M$ (i.e. $yC(x_0) = f(x_0)$ for at least one $x_0 \in \{f \neq 0\}$).

Remark

We exclude "trivial" implicants yC of f with $yC = f$ only for all $x \in \{f = 0\}$ since such implicants may be omitted in each representation

$$f = \max(I_1, ..., I_r)$$

of f by implicants (see Definition 5.5.1).
In the following the relation $f < g \quad (f, g : M \to \mathbb{R})$ always means $f \leq g$ and $f \neq g$.

Now we define prime implicants of f.

5.4.2 Definition

An implicant I of f is called *prime* implicant of f if there is no implicant I' of f with $I' > I$.
We remark that Definition 5.4.1 and Definition 5.4.2 are generalizations of Definition 2.2.2 and Definition 2.3.1.

To characterize prime implicants of f we need the following result.

5.4.3 Theorem

Let $yC, y'C'$ be implicants of f. Then the following are equivalent:

(a) $yC < y'C'$.
(b) $y = y', C < C'$.

Proof

Only (a) $\Rightarrow$ (b) is to prove. It follows from the implicant property of yC that there is an $x_0 \in M$ with $yC(x_0) = y = f(x_0)$. Now (a) yields $y'C'(x_0) = y' \geq yC(x_0) =$

y. And the implicant property of $y'C'$ implies $y' = y'C'(x_0) \leq f(x_0) = y$ and so $y' = y$. From (a) now we deduce $C < C'$. □

Theorem 5.4.3 yields the following characterization of prime implicants.

5.4.4 Theorem

Let yC be an implicant of f. Then the following are equivalent:

(a) yC is prime implicant of f.

(b) There is no implicant yC' of f with $C < C'$.

Now let us return to the monotone functions treated in Section 5.2 and characterized among other things by the unique representations of Theorem 5.2.17. We will show that in fact an isotone function has a unique representation as maximum of prime implicants, and this is the representation (5.2.18) as maximum of all its prime implicants.

Likewise an antitone function has a unique representation by prime implicants given by (5.2.19) as maximum of all its prime implicants.

First we state the set of all prime implicants of an isotone function.

5.4.5 Theorem

Denote by $C_p(f)$ the set of all prime implicants of f.

a) Let f be isotone. Then

$$C_p(f) = \{y1_{[z,a*]} : y \in \{y_1, ..., y_k\}, z \in G_{y*}\}, \tag{5.4.6}$$

(a^* and G_{y^*} as in Theorem 5.2.17).

b) Let f be antitone. Then

$$C_p(f) = \{y1_{[0,z]} : y \in \{y_1, ..., y_k\}, z \in G_y^*\}, \tag{5.4.7}$$

(G_y^* as in Theorem 5.2.17).

Proof

a) Let f be isotone.

1. Suppose $y \in \{y_1, ..., y_k\}, z \in G_{y*}$. We show that $y1_{[z,a^*]}$ is a prime implicant of f. From

$$y1_{[z,a^*]}(x) = \begin{cases} y & \text{for } z \leq x \leq a^* \\ 0 & \text{otherwise} \end{cases}$$

we obtain

$$y1_{[z,a^*]}(x)\begin{cases} = f(x) = y & \text{for } x = z \\ \leq f(x) & \text{otherwise.} \end{cases}$$

Thus $y1_{[z,a^*]}$ is an implicate of f. Assume $y1_{[z,a^*]}$ not to be a prime implicant of f. Then due to Theorem 5.4.4 there is an implicant yC' of f with $1_{[z,a^*]} < C'$ and $C' = 1_{\times P_i}$ where $\times P_i$ is a cube with $[z, a^*] \subset \times P_i$. Then $\times P_i$ contains a point $z' \in M$ with $z' < z$ implying $yC'(z') = y$ in contradiction to the supposition z to be a minimal point of $\{f = y\}$. Thus $y1_{[z,a^*]}$ is a prime implicant of f.

2. Suppose $y1_{\times P_i}$ to be a prime implicant of f. Let z be the minimal point of $\times P_i$ and z' the maximal point of $\times P_i$. Then from $y1_{\times P_i}(z) = y \leq f(z)$ it follows $y1_{\times P_i}(x) = y \leq f(x)$ for all $x \in [z, z']$ due to the isotony of f. Thus $\times P_i = [z, z']$ holds. Assume $z \notin \{f = y\}$. Then it follows $y1_{[z,z']}(x) = y < f(x)$ for all $x \in [z, z']$ and so $y1_{[z,z']}(x) = f(x)$ only for $f(x) = 0$ in contradiction to the definition of a prime implicant. Thus $z \in \{f = y\}$. Now the assumption $z \notin G_{y^*}$ or $z' < a^*$ implies that there is a $z'' \in G_{y^*}$ with $z'' \leq z$ so that $y1_{[z'',a^*]}$ is a prime implicant with $y1_{[z,z']} < y1_{[z'',a^*]}$. Then $y1_{[z,z']}$ cannot be a prime implicant. Thus $y1_{\times P_i}$ is of the proposed form.

b) The proof of b) is the same. □

Theorem 5.2.17 and Theorem 5.4.5 show that the representations (5.2.18) and (5.2.19) of monotone functions are the unique representations by prime implicants, namely by all their prime implicants. We may state the following result.

5.4.8 Theorem

The unique representation of a monotone function f by prime implicants is

$$f = \max_{i \in C_p(f)} I$$

as maximum of all its prime implicants.

Now let us consider general (not necessary monotone) discrete functions.

In the following we show how the implicants (prime implicants) of a discrete function f may be obtained in a simple way from the implicants (prime implicants) of the indicators $1_{\{f \geq y_i\}}$, $i \in N_k$. First we define some subsets of implicants (prime implicants) of f.

5.4.9 Definition

Let $\mathcal{C}(f)$ be the set of all implicants of f. For $i \in N_k$, let

(5.4.10) $$\mathcal{C}_{y_i}(f) := \{yC \in \mathcal{C}(f) : y = y_i\}$$

(set of all implicants of f of type y_iC),

and

(5.4.11) $$\mathcal{C}_{y_ip}(f) := \{yC \in \mathcal{C}_p(f) : y = y_i\}$$

(set of all prime implicants of f of type y_iC).

Obviously

(5.4.12) $$\mathcal{C}(f) = \dot{\bigcup_{i \in N_k}} \mathcal{C}_{y_i}(f),$$

(5.4.13) $$\mathcal{C}_p(f) = \dot{\bigcup_{i \in N_k}} \mathcal{C}_{y_ip}(f).$$

The following theorem yields the sets $\mathcal{C}_{y_i}(f), \mathcal{C}_{y_ip}(f), i \in N_k$.

5.4.14 Theorem

For $i \in N_k$, any y_iC is an implicant (prime implicant) of f if and only if C is an implicant (prime implicant) of $1_{\{f \geq y_i\}}$ but not of $1_{\{f > y_i\}}$ (if $i < k$) i.e.

(5.4.15) $$\mathcal{C}_{y_i}(f) = \{y_iC : C \in \mathcal{C}(\{f \geq y_i\}) \setminus \mathcal{C}(\{f > y_i\})\},$$

(5.4.16) $$\mathcal{C}_{y_ip}(f) = \{y_iC : C \in \mathcal{C}_p(\{f \geq y_i\}) \setminus \mathcal{C}_p(\{f > y_i\})\},$$

(where $\mathcal{C}(\{f > y_k\}) = \mathcal{C}_p(\{f > y_k\}) = \phi$).

Proof

1. Let y_iC be an implicant of f. Then $C = 0$ for each $x \in \{f < y_i\}$, thus C is an implicant of $1_{\{f \geq y_i\}}$. Further (if $i < k$) $y_iC = y_i = f$ and so $C = 1$ for some $x \in \{f = y_i\}$, i.e. $x \notin \{f > y_i\}$, thus C is not an implicant of $1_{\{f > y_i\}}$.

2. Let C be an implicant of $1_{\{f\geq y_i\}}$ but not of $1_{\{f>y_i\}}$ (if $i<k$). Then $y_iC = y_i = f$ for some $x \in \{f = y_i\}$, further $y_iC \leq y_i \leq f$ for $x \in \{f \geq y_i\}$ and $y_iC = 0 \leq f$ for $x \in \{f < y_i\}$, thus y_iC is an implicant of f.
3. Let y_iC be a prime implicant of f. Then C is an implicant of $1_{\{f\geq y_i\}}$ but not an implicant of $1_{\{f>y_i\}}$ (if $i<k$) due to 1, and so not a prime implicant of $1_{\{f>y_i\}}$. Let $C < C'$. Then by supposition y_iC' is not an implicant of f, thus $y_iC' = y_i$ and so $C' = 1$ for some $x \in \{f < y_i\}$, i.e. C' is not an implicant of $1_{\{f\geq y_i\}}$ and so C is a prime implicant of $1_{\{f\geq y_i\}}$.
4. Let C be a prime implicant of $1_{\{f\geq y_i\}}$ but not of $1_{\{f>y_i\}}$ (if $i<k$). Then y_iC is an implicant f due to 2. Let $C < C'$. Then C' is not an implicant of $1_{\{f\geq y_i\}}$ and so $y_iC' = y_i > f$ for some $x \in \{f < y_i\}$. Thus y_iC is a prime implicant of f. □

Now $C(f)$ and $C_p(f)$ are given by (5.4.12), (5.4.13), (5.4.15), (5.4.16).

5.5 Representations by Implicants (Prime Implicants)

First we state the definitions of the several representations of f by implicants (see also the corresponding definitions of Section 2.4).

5.5.1 Definition

If $I_1, \dots, I_r$ are implicants (prime implicants) of f with

(5.5.2) $$f = \max_{\rho \in N_r} I_\rho$$

then we call (5.5.2) a *representation of f by implicants (prime implicants).*

If moreover

$$f \neq \max_{\rho \in K} I_\rho$$

for each $K \subset N_r$ then we call (5.5.2) a *reduced* representation of f by implicants (prime implicants).

If finally $r' \geq r$ for each representation

$$f = \max_{\rho \in N_{r'}} I'_\rho$$

of f by implicants (prime implicants) then we call (5.5.2) a *minimal* representation of f by implicants (prime implicants).

5.5.3 Definition

A prime implicant I of f is called *essential* if $I \in \{I_1, ..., I_r\}$ for each representation (5.15) of f by prime implicants.

Generalizing Definition 2.4.12 we may characterize the several sets of representations (reduced representations) of f by implicants (prime implicants) in the following way.

5.5.4 Definition

We define

$$\mathcal{I}(f) := \{B \subseteq \mathcal{C}(f) : \max_{I \in B} I = f\},$$
$$\mathcal{I}_p(f) := \{B \subseteq \mathcal{C}_p(f) : \max_{I \in B} I = f\},$$
$$\mathcal{R}(f) := \{B \in \mathcal{I}(f) : \max_{i \in B} \neq f \text{ for } K \subset B\},$$
$$\mathcal{R}_p(f) := \{B \in \mathcal{I}_p(f) : \max_{I \in K} I \neq f \text{ for } K \subset B\} = \mathcal{R}(f) \cap \mathcal{I}_p(f).$$

Obviously this means:

$B \in \mathcal{J}(f) \Leftrightarrow f = \max_{I \in B} I$ is a representation of f by **implicants**;

$B \in \mathcal{I}_p(f) \Leftrightarrow f = \max_{I \in B} I$ is a representation of f by **prime implicants**;

$B \in \mathcal{R}(f) \Leftrightarrow f = \max_{I \in B} I$ is a **reduced** representation of f by **implicants**;

$B \in \mathcal{R}_p(f) \Leftrightarrow f = \max_{I \in B} I$ is a **reduced** representation of f by **prime implicants**.

Again $\mathcal{R}_p(f)$ is not empty. The order relation of the four sets is given by the following figure.

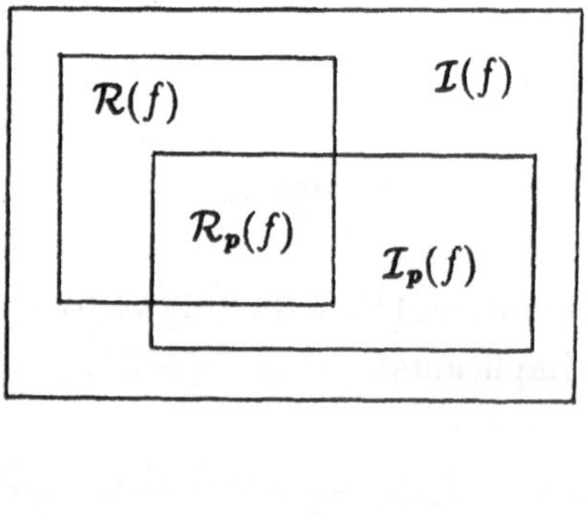

The next result is a consequence of Definition 5.4.1 and Definition 5.5.4.

5.5.5 Proposition

Let $B \subset \mathcal{C}(f)$ $(\mathcal{C}_p(f))$. Then $B \in \mathcal{I}(f)$ $(\mathcal{I}_p(f))$ if and only if for each $x \in \{f > 0\}$ there is at least one $I \in B$ with $I(x) = f(x)$.

Remark

Since for each $x \in \{f > 0\}$ there is at least one $I \in \mathcal{C}(f)$ $(\mathcal{C}_p(f))$ with $I(x) = f(x)$, we have

$$\max_{I \in \mathcal{C}(f)} I = \max_{I \in \mathcal{C}_p(f)} I = f.$$

Thus the function f equals the maximum of all its implicants (prime implicants). (See also Corollary 2.4.10.)

From Definition 2.4.4 and Definition 2.4.14 we obtain

(5.5.6) $\mathcal{R}(f) = R(\mathcal{I}(f))$,
(5.5.7) $\mathcal{R}_p(f) = R(\mathcal{I}_p(f))$.

The next result is a generalization of Theorem 2.4.15 to discrete functions.

5.5.8 Theorem

For $i \in N_k$, let $\{f = y_i\} = \{\alpha_{i1}, ..., \alpha_{iq_i}\} (\subset M)$.
Define for $i \in N_k$, $\kappa \in N_q$

$$\mathcal{M}_{i\kappa}(f) := \{I \in \mathcal{C}_{y_i}(f) : I(\alpha_{i\kappa}) = y_i\}$$

$(\mathcal{C}_{y_i}$ defined by (5.4.10)),

$$\mathcal{M}_{i\kappa p}(f) := \{I \in \mathcal{C}_{y_i p}(f) : I(\alpha_{i\kappa}) = y_i\}$$

$(\mathcal{C}_{y_i p}$ defined by (5.4.11)),

$$\mathcal{I}_i(f) := \mathcal{S}(\mathcal{M}_{i1}(f), ..., \mathcal{M}_{iq_i}(f)),$$

$$\mathcal{I}_{ip}(f) := \mathcal{S}(\mathcal{M}_{i1p}(f), ..., \mathcal{M}_{iq_ip}(f)),$$

($\mathcal{S}$ defined by 2.4.13, see also Theorem 2.4.15).

Then

(5.5.9) $\mathcal{I}(f) = \{ \bigcup_{i \in N_k} B_i : B_i \in \mathcal{I}_i(f) \text{ for } i \in N_k \}$,

(5.5.10) $\mathcal{I}_p(f) = \{ \bigcup_{i \in N_k} B_i : B_i \in \mathcal{I}_{ip}(f) \text{ for } i \in N_k \}$,

(5.5.11) $\mathcal{R}(f) = \{ \bigcup_{i \in N_k} B_i : B_i \in R(\mathcal{I}_i(f)) \text{ for } i \in N_k \}$,

(5.5.12) $\mathcal{R}_p(f) = \{ \bigcup_{i \in N_k} B_i : B_i \in R(\mathcal{I}_{ip}(f)) \text{ for } i \in N_k \}$.

Remark

For $i \in N_k$, $\kappa \in N_{q_i}$, we have

$$\mathcal{M}_{i\kappa}(f) \neq \emptyset \ , \ \mathcal{M}_{i\kappa p}(f) \neq \emptyset \ ,$$
$$\bigcup_{\kappa \in N_{q_i}} \mathcal{M}_{i\kappa}(f) = \mathcal{C}_{y_i}(f) \ , \ \bigcup_{\kappa \in N_{q_i}} \mathcal{M}_{i\kappa p}(f) = \mathcal{C}_{y_i p}(f) \ .$$

Proof of (5.5.9)

1. Suppose $B \in \mathcal{I}(f)$. From (5.4.12) it follows

$$B = B \cap \mathcal{C}(f) = \bigcup_{i \in N_k} B \cap \mathcal{C}_{y_i}(f) = \bigcup_{i \in N_k} B_i$$

with $B_i := B \cap \mathcal{C}_{y_i}(f) \subseteq \mathcal{C}_{y_i}(f)$. We have to show that $B_i \in \mathcal{I}_i(f)$ for each $i \in N_k$.
Assume $B_i \notin \mathcal{I}_i(f)$ for some $i \in N_k$. Then by definition of $\mathcal{I}_i(f)$ (see also Definition 2.4.13) there is a $\kappa \in N_{q_i}$ with $B_i \cap \mathcal{M}_{i\kappa}(f) = \emptyset$ and so $I(\alpha_{i\kappa}) \neq y_\rho$ for each $I \in B_i$. Since $I(\alpha_{i\kappa}) \neq y_i$ also for each $I \in B_{i'}$ with $i' \in N_k \setminus \{i\}$, it follows $\max_{I \in B} I(\alpha_{i\kappa}) \neq f(\alpha_{i\kappa})$ and so $B \notin \mathcal{I}(f)$ in contradiction to the supposition. Thus $B_i \in \mathcal{I}_i(f)$ for each $i \in N_k$.

2. Suppose $B \in \{ \bigcup_{i \in N_k} B_i : B_i \in \mathcal{I}_i(f) \text{ for } i \in N_k \}$ and $x \in \{f > 0\}$. Then $x = a_{i\kappa}$ for some $a_{i\kappa} \in \{f = y_i\}, i \in N_k$ and $B_i \cap \mathcal{M}_{i\kappa} \neq \emptyset$ by definition of $\mathcal{I}_i(f)$. Thus $I(x) = y_i = f(x)$ for some $I \in B_i \subset B$. Thus $B \in \mathcal{I}(f)$ due to Proposition (5.5.5). □

The proof of (5.5.10) is of the same kind.

Proof of (5.5.11) and (5.5.12)

The proof follows from (5.5.6), (5.5.7) and the following lemma.

5.5.13 Lemma

Let $A_1, ..., A_k$ be finite sets, $A_i \cap A_j = \emptyset$ for $i \neq j$ and $\mathcal{A}_1 \subseteq \mathcal{P}(A_1), ..., \mathcal{A}_k \subseteq \mathcal{P}(A_k)$. Define

$$\mathcal{A} := \{ \dot{\bigcup_{i \in N_k}} B_i : B_i \in \mathcal{A}_i \text{ for } i \in N_k \}.$$

$$\mathcal{A}' := \{ \dot{\bigcup_{i \in N_k}} B_i : B_i \in R(\mathcal{A}_i) \text{ for } i \in N_k \}.$$

Then

$$R(\mathcal{A}) = \mathcal{A}'.$$

Proof

a) Suppose $B := \dot{\bigcup_{i \in N_k}} B'_i \in R(\mathcal{A})$ and assume $B \notin \mathcal{A}'$. Then $B_i \notin R(\mathcal{A}_{i'})$ for at least one $i' \in N_k$ and so $B''_{i'} \subset B'_{i'}$ for some $B''_{i'} \in \mathcal{A}_{i'}$. But then $B' = B''_{i'} \cup \dot{\bigcup_{i \neq i'}} B'_i \subset B, B' \not\subset \mathcal{A}$ which implies $B \notin R(\mathcal{A})$ in contradiction to the supposition. Thus $B \in \mathcal{A}'$.

b) Suppose $B := \dot{\bigcup_{i \in N_k}} B'_i \in \mathcal{A}'$. Then $B \in \mathcal{A}$. Assume $B \notin R(\mathcal{A})$. Then there is a $B' := \dot{\bigcup_{i \in N_k}} B''_i \in \mathcal{A}$ with $B' \subset B$ and so $B''_{i'} \subset B_{i'}, B''_{i'} \in \mathcal{A}_{i'}$ for some $i' \in N_k$, thus $B'_{i'} \notin R(\mathcal{A}_{i'})$ in contradiction to the supposition. Thus $B \in R(\mathcal{A})$. □

Résumé

Comprising the results of this section we obtain the following method to construct stepwise the sets $\mathcal{I}(f), \mathcal{I}_p(f), \mathcal{R}(f), \mathcal{R}_p(f)$, where the last set $\mathcal{R}_p(f)$ yields immediately all reduced representations of the discrete function f with $f(M) = \{0, y_1, ..., y_k\}$ by prime implicants. It may suffice to consider $\mathcal{I}_p(f)$ and $\mathcal{R}_p(f)$. The other case is obvious.

Step 1

In the first step we determine the prime indicator sets $\mathcal{C}_p(f \geq y_i\}, i \in N_k$ by the methods of Chapter 2.

Step 2

By (5.4.16) of Theorem 5.4.14 we obtain the sets $\mathcal{C}_{y_i p}(f)$ of all prime implicants of type $y_i C$ taking the values y_i and 0, $i \in N_k$.

Step 3

For each $i \in N_k$ and each point $\alpha_{i\kappa} \in \{f = y_i\} = \{\alpha_{i1}, ..., \alpha_{iq_i}\}$, we determine the set $\mathcal{M}_{i\kappa p}(f)$ according to Theorem 5.5.8.

Step 4

Now we determine $\mathcal{I}_{ip}(f), i \in N_k$ according to Theorem 5.5.8.

Step 5

We obtain $\mathcal{I}_p(f)$ by (5.5.10).

Step 6

The last step (5.5.12) of Theorem 5.5.8 yields the set $\mathcal{R}_p(f)$, i.e all sets B of prime implicates of f such that

$$f = \max_{I \in B} I$$

is a reduced representation of f by prime implicates.

Obviously in the last step we shall use the reduction methods of Chapter 4.

Supplement to Theorem 5.5.8

Let for $i \in N_k$ the sets $C_{y_i}(f)$ of all implicants of f of type y_iC be given by

$$C_{y_i}(f) = \{I_{i1}, ..., I_{ir_i}\}.$$

Then

(a) $$f = \max_{I \in B} I \quad (\text{with } B \subseteq C(f)$$

holds if and only if B contains for each $i \in N_k$ and for each $x \in \{f = y_i\}$ at least one $I_{i\rho}$ with $\rho \in N_{r_i}$ and $I_{i\rho}(x) = y_i$. This suggests to decompose $\{f = y_i\}$ in disjoint subsets in the following way. For each $N \subseteq N_{r_i}$, define M_N by

(b) $$M_N := \left\{ x \in \{f = y_i\} : I_{i\rho}(x) = \begin{cases} y_i & \text{for } \rho \in N \\ 0 & \text{otherwise} \end{cases} \right\}.$$

Then $M_N \cap M_{N'} = \emptyset$ for $N \neq N'$ and $\bigcup_{N \subseteq N_{r_i}} M_N = \{f = y_i\}$. Obviously for (a) it is necessary and sufficient that B contains to each nonempty set M_N at least one $I_{i\rho}$ with $\rho \in N$, and in most cases the number of the nonempty sets will be smaller than $|\{f = y_i\}|$. This is the idea of the next statement yielding the sets $\mathcal{I}_i(f)$ and $\mathcal{I}_{ip}(f)$ of Theorem 5.5.8 in a simpler form.

5.5.8' Supplement

For $i \in N_k$, let $\mathcal{C}_{y_i}(f)$ and $\mathcal{C}_{y_ip}(f)$ be given by

$$\mathcal{C}_{y_i}(f) = \{I_{i1}, ..., I_{ir_i}\},$$
$$\mathcal{C}_{y_ip}(f) = \{I'_{i1}, ..., I'_{ir'_i}\},$$

further $\mathcal{I}_i(f)$ and $\mathcal{I}_{ip}(f)$ as in Theorem 5.5.8. Let M_N be given by (b) and define

$$M'_N := \left\{x \in \{f = y_i\} : I'_{i\rho}(x) = \begin{cases} y_i & \text{for } \rho \in N \\ 0 & \text{otherwise} \end{cases}\right\}.$$

Write

$\{N \subseteq N_{r_i} : M_N \neq \emptyset\} = \{N_{i1}, ..., N_{is_i}\}$,
$\{N \subseteq N'_{r_i} : M'_N \neq \emptyset\} = \{N'_{i1}, ..., N'_{is'_i}\}$,
$\mathcal{N}_{ij}(f) := \{I_{i\rho} \in \mathcal{C}_{y_i}(f) : \rho \in N_{ij}\} = \{I \in \mathcal{C}_{y_i}(f) : I(x) = y_i \text{ for } x \in M_{N_{ij}}\}$,
$\qquad j \in N_{s_i}$,
$\mathcal{N}_{ijp}(f) := \{I'_{i\rho} \in \mathcal{C}_{y_ip}(f) : \rho \in N'_{ij}\} = \{I \in \mathcal{C}_{y_ip}(f) : I(x) = y_i \text{ for } x \in M'_{N_{ij}}\}$,
$\qquad j \in N_{s'_i}$.

Then (conf. Theorem 5.5.8) for $i \in N_k$, we have

$$\{\mathcal{N}_{i1}(f), ..., \mathcal{N}_{is_i}(f)\} = \bigcup_{j \in N_{q_i}} \{\mathcal{M}_{ij}(f)\},$$
$$\{\mathcal{N}_{i1p}(f), ..., \mathcal{N}_{is'_ip}(f)\} = \bigcup_{j \in N_{q_i}} \{\mathcal{M}_{ijp}(f)\}$$

and

$$\mathcal{I}_i(f) = \mathcal{S}(\mathcal{N}_{i1}(f), ..., \mathcal{N}_{is_i}(f)),$$
$$\mathcal{I}_{ip}(f) = \mathcal{S}(\mathcal{N}_{i1p}(f), ..., \mathcal{N}_{is'_ip}(f)) .$$

Proof

The sets M_N with $N \in \{N_{i1}, ..., N_{is_i}\}$ are a partition of $\{f = y_i\}$, i.e.

$$\{f = y_i\} = \dot{\bigcup_{j \in N_{s_i}}} M_{N_{ij}}.$$

From the definition of $\mathcal{N}_{ij}(f)$ we obtain $\mathcal{N}_{ij}(f) \neq \mathcal{N}_{ij'}(f)$ for $j \neq j'$ and (see Theorem 5.5.8) $\mathcal{M}_{i\kappa}(f) = \mathcal{N}_{ij}(f) = \{I_{i\rho} \in \mathcal{C}_{y_i}(f) : \rho \in N_{ij}\}$ for all $\alpha_{i\kappa} \in M_{N_{ij}}$ and thus

$$\{\mathcal{N}_{i1}(f), ..., \mathcal{N}_{is_i}(f)\} = \bigcup_{j \in N_{q_i}} \{\mathcal{M}_{ij}(f)\}$$

implying

$$\mathcal{I}_i(f) := \mathcal{S}(\mathcal{M}_{i1}(f), ..., \mathcal{M}_{iq_i}(f)) = \mathcal{S}(\mathcal{N}_{i1}(f), ..., \mathcal{N}_{is_i}(f)\}.$$

The proof in case of prime implicates is the same. □

Remark

The set $\{\mathcal{N}_{i1}(f), ..., \mathcal{N}_{is_i}(f)\}$ contains all **different** elements of the sequence $\mathcal{M}_{i1}(f), ..., \mathcal{M}_{iq_i}(f)$.

Write $I_{ij} = y_i C_{ij}$ for $j \in N_{r_i}$ $(I_{ij} \in \mathcal{C}_{y_i}(f))$. Then

$$M_N = \{x \in \{f = y_i\} : \prod_{j \in N} C_{ij} \prod_{j \in N_{ri} \setminus N} (1 - C_{ij}) = 1\}.$$

With $C_{ij} = 1_{P_{ij}}$ $(P_{ij}$ a cube), we obtain

$$M_N = \{f = y_i\} \cap \bigcap_{j \in N} P_{ij} \cap \bigcap_{j \in N_{r_i} \setminus N} \overline{P_{ij}}.$$

The corresponding holds in case of prime implicants.

We still note that in case $k = 1$ our supplement 5.8' yields a corresponding supplement to Theorem 2.4.15.

5.6 Implicates (Prime Implicates) of Discrete Functions

Now we give another kind of representations of f as a minimum of "implicates". First we define implicates and prime implicates of f.

5.6.1 Definition

Let F be an anticube indicator of M (Definition 3.1.1) and $\overline{y_i} := y_k - y_i$ for $i \in \{0, ..., k-1\}$ (with $y_0 := 0$). Then for $i \in \{0, ..., k-1\}$

$$I^*(y_i, F) := \max(y_i, y_k F) = y_i + \overline{y_i} F = \begin{cases} y_k & \text{for } F = 1 \\ y_i & \text{for } F = 0 \end{cases} \tag{5.6.2}$$

is called a (nontrivial) *implicate* of f if $y_i + \overline{y_i} F \geq f$ and $y_i \overline{y_i} F(x_0) = f(x_0) = y_i$ for at least one $x_0 \in \{f = y_i\}$.

Remark

We exclude "trivial" implicates I^* of f with $y_i + \overline{y_i}F = f$ for $x \in \{f = y_k\}$ since such implicates may be omitted in each representation

$$f = \min(I_1^*, ..., I_r^*)$$

by implicates.

5.6.3 Definition

An implicate I^* of f is called *prime* implicate if there is no other implicate $I^{*'}$ of f with $I^{*'} \leq I^*$.

We may regard Definition 5.6.1 and Definition 5.6.3 as generalizations of Definition 3.2.5 and Definition 3.3.1 concerning implicates and prime implicates of indicators.

5.6.4 Theorem

Let $y_i + \overline{y_i}F$, $y_j + \overline{y_j}F'$ be implicates of f. Then the following are equivalent:

(a) $y_j + \overline{y_j}f' < y_i + \overline{y_i}F$.
(b) $i = j$ and $F' < F$.

Proof

Only (a)$\Longrightarrow$ (b) has to be proved. From the implicate property of $y_i + \overline{y_i}F$ it follows $y_i + \overline{y_i}F(x_0) = y_i = f(x_0)$ for some $x_0 \in \{f = y_i\}$. Now (a) yields $y_j + \overline{y_j}F'(x_0) = y_j \leq y_i + \overline{y_i}F(x_o) = y_i$. Further the implicate property of $y_j + \overline{y_j}F'$ implies $y_j = y_j + \overline{y_j}F'(x_0) \geq f(x_0) = y_i$ and so $y_j = y_i$, i.e. $i = j$. From (a) now we deduce $F' < F$. □

From Theorem 4.6.5 we obtain the following characterization of prime implicates.

5.6.5 Theorem

Let $y_i + \overline{y_i}F$ be an implicate of f. Then the following are equivalent:

(a) $y_i + \overline{y_i}F$ is a prime implicate of f.
(b) There is no implicate $y_i + \overline{y_i}F'$ of f with $F' < F$.

The next statement gives the connection between the implicants (prime implicants) of a discrete function f and the implicates (prime implicates) of the complementary discrete function $\overline{f} := y_k - f$.

5.6.6 Theorem

Let f be a discrete function, $\overline{f} := y_k - f$ the complementary discrete function (with $\overline{f}(M) = \{0, y_k - y_{k-1}, \ldots, y_k\}$). Let C be a cube indicator, F an anticube indicator and $C + F = 1$. Then for each $y \in \{y_1, \ldots, y_k\}$, the following are equivalent:

(a) $y_i C$ is an implicant (prime implicant) of f.

(b) $\overline{y_i} + y_i F$ with $\overline{y_i} := y_k - y_i \in \{0, y_k - y_{k-1}, \ldots, y_k - y_1\}$ is an implicate (prime implicate) of $\overline{f}$.

We can express this also in the following way. Let $\mathcal{F}(f)$ ($\mathcal{F}_p(f)$) be the set of all implicates (prime implicates) of f. Then

$$(5.6.7) \qquad C(f) = \{I : y_k - I \in \mathcal{F}(\overline{f})\} = \{y_k - I^* : I^* \in \mathcal{F}(\overline{f})\},$$

$$(5.6.8) \qquad C_p(f) = \{I : y_k - I \in \mathcal{F}_p(\overline{f})\} = \{y_k - I^* : I^* \in \mathcal{F}_p(\overline{f})\}.$$

This means:
I is an implicant (prime implicant) of f if and only if $y_k - I$ is an implicate (prime implicate) of $\overline{f}$, i.e. an implicate (prime implicate) I^* of $\overline{f}$.

Proof

It holds $y_k - y_i C = y_k - y_i(1 - F) = y_k - y_i + y_i F = \overline{y_i} + y_i F$.
Let $I := yC$, $I^* := y_k - I$. Then the equivalence

$$(I \leq f \text{ , } I(x_0) = f(x_0) \text{ for } x_0 \in \{f \neq 0\})$$
$$\Leftrightarrow (I^* > \overline{f} \text{ , } I^*(x_0) = \overline{f}(x_0) \text{ for } x_0 \in \{\overline{f} \neq y_k\})$$

yields the proposition for implicants of f (implicates of $\overline{f}$).

Now suppose I to be a prime implicant of f. Then $I^* := y_k - I$ is an implicate of $\overline{f}$. Assume I^* not to be a *prime* implicate of $\overline{f}$. Then there is an implicate I'^* of $\overline{f}$ with $I'^* < I^*$. Then $I' := y_k - I'^*$ is an implicant of f with $I' = y_k - I'^* > y_k - I^* = I$ in contradiction to the supposition. Thus I^* is prime implicate of $\overline{f}$.

Finally suppose I^* to be a prime implicate of $\overline{f}$. Then $I := y_k - I^*$ is an implicant of f. Assume I not to be a *prime* implicant of f. Then there is an implicant I' of f with $I < I'$. Then $I'^* = y_k - I'$ is an implicate of $\overline{f}$ with $I'^* = y_k - I' < y_k - I = I^*$ in contradiction to the supposition. Thus I is prime implicant of f. □

Exchanging f and $\overline{f}$ we deduce from Theorem 5.6.6 the corresponding statement concerning the implicates (prime implicates) of f and the implicants (prime implicants) of $\overline{f}$.

5.6.9 Theorem

Let $f, \overline{f}$, C and F be given as in Theorem 5.6.6. Then for $i \in \{0, ..., k-1\}$, the following are equivalent:

(a) $y_i + \overline{y_i}F$ is an implicate (prime implicate) of f.

(b) $y_k - (y_i + \overline{y_i}F) = \overline{y_i}C$ with $\overline{y}_i = y_k - y_i \in \{y_k - y_{k-1}, ..., y_k\}$ is an implicant (prime implicant) of $\overline{f}$.

This again means:

$$\mathcal{F}(f) = \{I^* : y_k - I^* \in \mathcal{C}(\overline{f})\} = \{y_k - I : I \in \mathcal{C}(\overline{f})\}, \tag{5.6.10}$$

$$\mathcal{F}_p(f) = \{I^* : y_k - I^* \in \mathcal{C}_p(\overline{f})\} = \{y_k - I : I \in \mathcal{C}_p(\overline{f})\}. \tag{5.6.11}$$

Thus we may state:
I^* is an implicate (prime implicate) of f if and only if $y_k - I^*$ is an implicant (prime implicant of $\overline{f}$, i.e. if and only if I^* is the complement $y_k - I$ of an implicant (prime implicant) I of $\overline{f}$.

In Section 5.4 we stated in Theorem 5.4.5 the particularly simple form of all prime implicants of monotone discrete functions. Moreover in Theorem 5.4.8 we showed that monotone functions have a unique representation as the maximum of all their prime implicants.

Now we may use Theorem 5.6.9 to obtain the corresponding results about prime implicates of monotone functions and the unique representation of monotone functions as the minimum of all their prime implicates.

5.6.12 Theorem

a) Let f be isotone and G_y^* the set of all maximal points of $\{f = y\}, y \in \{0, ..., y_{k-1}\}$. Then the set $\mathcal{F}_p(f)$ of all prime implicates of f is given by

$$\mathcal{F}_p(f) = \{y + \overline{y}1_{\overline{[0,z]}} : y \in \{0, ..., y_{k-1}\}, z \in G_y^*\}. \tag{5.6.13}$$

b) Let f be antitone and G_{y^*} the set of all minimal points of $\{f = y\}, y \in \{0, ..., y_{k-1}\}$. Then $\mathcal{F}_p(f)$ is given by

$$\mathcal{F}_p(f) = \{y + \overline{y}1_{\overline{[z,a^*]}} : y \in \{0, ..., y_{k-1}\}, z \in G_{y^*}\}. \tag{5.6.14}$$

Proof

a) Let f be isotone and so $\overline{f} = y_k - f$ antitone.
1. Suppose $y \in \{0, ..., y_{k-1}\}$, $z \in G_y^*$. We show that $y + \overline{y}1_{\overline{[0,z]}}$ is a prime implicant of f. From $\overline{y} \in \{y_k - y_{k-1}, ..., y_k\}$, $z \in G_y^*$, and $\{f = y\} = \{\overline{f} = \overline{y}\}$ it follows by Theorem 5.4.5, b) that $\overline{y}1_{[0,z]}$ is a prime implicant of $\overline{f}$. Now from Theorem 5.6.9 it follows that $y + \overline{y}F$ with $F = 1 - 1_{[0,z]} = 1_{\overline{[0,z]}}$ is a prime implicate of f.
2. Let $y + \overline{y}F$ be prime implicate of f. Then $\overline{y}C$ with $C = 1 - F$ is prime implicant of $\overline{f}$ and so $C = 1_{[0,z]}$ with $z \in G_y^*$ (since $\{\overline{f} = \overline{y}\} = \{f = y\}$). Thus $F = 1 - C = 1_{\overline{[0,z]}}$ with $z \in G_y^*$.

b) The proof of b) is the same.

Now from Theorem 5.6.12 and Theorem 5.2.20 it follows that the representations (5.2.21) and (5.2.22) are the unique representations of monotone discrete functions, given by the minimum of all its prime implicates, i.e. we may state the following result.

5.6.15 Theorem

The unique representation of a monotone function f by prime implicates is

$$f = \min_{I^* \in \mathcal{F}_p(f)} I^*$$

as minimum of all its prime implicates.

Let us return again to general (not necessarily monotone) discrete functions. In Section 5.4, Theorem 5.4.14 we showed how the implicants (prime implicants) of f can be obtained with the aid of the implicants (prime implicants) of the indicators $1_{\{f \geq y_i\}}, i \in N_k$.

Now we will state the corresponding result concerning the **implicates** of f.

5.6.16 Definition

For each $i \in \{0, ..., k-1\}$ ($y_0 := 0$, $\mathcal{F}(f)$ and $\mathcal{F}_p(f)$ as in Theorem 5.6.6), we define

$$\mathcal{F}_{y_i}(f) := \{y + \overline{y}F \in \mathcal{F}(f) : y = y_i\} \tag{5.6.17}$$

(set of all implicates of f of type $y_i + \overline{y_i}F$),

$$\mathcal{F}_{y_ip}(f) := \{y + \overline{y}F \in \mathcal{F}_p(f) : y = y_i\} \tag{5.6.18}$$

(set of all prime implicates of f of type $y_i + \overline{y_i}F$).

Obviously

$$\mathcal{F}(f) = \dot{\bigcup_{i=0}^{k-1}} \mathcal{F}_{y_i}(f), \tag{5.6.19}$$

$$\mathcal{F}_p(f) = \dot{\bigcup_{i=0}^{k-1}} \mathcal{F}_{y_ip}(f). \tag{5.6.20}$$

5.6.21 Theorem

Any $y_i + \overline{y_i}F$ is an implicate (prime implicate) of f if and only if F is an implicate (prime implicate) of $1_{\{f>y_i\}}$ but not an implicate (prime implicate) of $1_{\{f\geq y_i\}}$ (if $i > 0$), i.e. for $i \in \{0, ..., k-1\}$, it holds

$$\mathcal{F}_i(f) = \{y_i + \overline{y_i}F : F \in \mathcal{F}(\{f > y_i\}) \setminus \mathcal{F}(\{f \geq y_i\})\}, \tag{5.6.22}$$

$$\mathcal{F}_{ip}(f) = \{y_i + \overline{y_i}F : F \in \mathcal{F}_p(\{f > y_i\}) \setminus \mathcal{F}_p(\{f \geq y_i\})\}, \tag{5.6.23}$$

(with $\mathcal{F}(\{f \geq 0\}) = \mathcal{F}_p(\{f \geq 0\}) = \emptyset$).

Proof

Let $C := 1 - F$. Then from Theorem 5.6.9, Theorem 5.4.14 and Theorem 3.2.5 we deduce that the following are equivalent:

(a) $y_i + \overline{y_i}F$ is implicate (prime implicate) of f.

(b) $\overline{y_i}C$ is implicant (prime implicant) of $\overline{f}$.

(c) C is implicant (prime implicant) of $1_{\{\overline{f}\geq\overline{y}_i\}} = 1_{\{f\leq y_i\}}$ but not implicant (prime implicant of $1_{\{\overline{f}>\overline{y}_i\}} = 1_{\{f<y_i\}}$.

(d) F is implicate (prime implicate) of $1_{\{f<y_i\}}$ but not implicate (prime implicate) of $1_{\{f\geq y_i\}}$. □

Now we obtain $\mathcal{F}$ and $\mathcal{F}_p(f)$ by (5.6.19), (5.6.20), (5.6.22) and (5.6.23).

5.7 Representations by Implicates (Prime Implicates)

5.7.1 Definition

If $I_1^*, ..., I_r^*$ are implicates (prime implicates) of f with

$$f = \min_{\rho \in N_r} I_\rho^* \tag{5.7.2}$$

then we call (5.7.2) a *representation of f by implicates (prime implicates).*

If moreover

$$f \neq \min_{\rho \in K} I_\rho^*$$

for each $K \subset N_r$ then we call (5.7.2) a *reduced* representation of f by implicates (prime implicates).

If finally $r' \geq r$ for each representation

$$f = \max_{\rho \in N_{r'}} I_\rho'^*$$

of f by implicates (prime implicates) then we call (5.7.2) a *minimal* representation of f by implicates (prime implicates).

5.7.3 Definition

A prime implicate I^* of f is called *essential* if $I^* \in \{I_1^*, ..., I_r^*\}$ for each representation (5.7.2) of f by prime implicates.

First we show how to gain (reduced minimal) representations of f by implicates (prime implicates) immediately from the corresponding implicant representations of $\overline{f}$.

5.7.4 Theorem

The discrete function f has the (reduced, minimal) representation

$$f = \min_{\rho \in N_r} I^* \tag{5.7.5}$$

by the implicates (prime implicates) $I_1^*, ..., I_r^*$ of f if and only if $\overline{f} = y_k - f$ has the (reduced, minimal) representation

$$\overline{f} = \max_{\rho \in N_r} I_\rho \tag{5.7.6}$$

by the implicants (prime implicants) $I_1 = y_k - I_1^*, ..., I_r = y_k - I_r^*$ of $\overline{f}$.

Proof

1. a) Suppose (5.7.5) to be a representation of f by the implicates (prime implicates) $I_1^*, ..., I_r^*$. Then due to Theorem 5.6.9 $I_1, ..., I_r$ are implicants (prime implicants) of $\overline{f}$. Now

$$\max_{\rho \in N_r} I_\rho = \max_{\rho \in N_r}(y_k - I_\rho^*) = y_k - \min_{\rho \in N_r} I_\rho^* = y_k - f = \overline{f}$$

implies that (5.7.6) is a representation of $\overline{f}$ by implicants (prime implicants).

b) Suppose (5.7.6) to be a representation of $\overline{f}$ by the implicants (prime implicants) $I_1, ..., .I_r$. Then due to Theorem 5.6.9 $I_1^*, ..., I_r^*$ are implicates (prime implicates) of f. Now it follows from

$$\min_{\rho \in N_r} I_\rho^* = \min_{\rho \in N_r}(y_k - I_\rho) = y_k - \max_{\rho \in N_r} I_\rho = y_k - \overline{f} = f,$$

that (5.7.5) is a representation of f by implicates (prime implicates).

2. a) Suppose (5.7.5) to be reduced. Assume (5.7.6) not to be reduced, i.e.

$$\overline{f} = \max_{\rho \in K} I_\rho \quad \text{with} \quad K \subset N_r.$$

Then due to 1. b)

$$f = \min_{\rho \in K} I_\rho^*$$

in contradiction to the supposition. Thus (5.7.6) is reduced.

b) In the same way we obtain that (5.7.5) is reduced if (5.7.6) is reduced.

3. a) Suppose (5.7.5) to be minimal. Assume (5.7.6) not to be minimal, i.e.

$$\overline{f} = \max_{\rho \in N_{r'}} I'_\rho \quad \text{with } r' < r$$

is a representation of $\overline{b}$ by implicates (prime implicates). Then due to 1. b)

$$f = \min_{\rho \in N_{r'}} I_\rho^{*'} \text{ with } r' < r \text{ and } I_\rho^{*'} = y_k - I'_\rho \text{ for } \rho \in N_{r'}$$

is a representation of f by implicates (prime implicates) in contradiction to the supposition (5.7.6) to be minimal. Thus (5.7.6) is minimal.

b) In the same way it follows that (5.7.5) is minimal if (5.7.6) is minimal. □

Obviously Theorem 5.7.4 allows us to gain all representations of f by implicates from the representations of the complementary function $\overline{f}$ by implicants (and vice versa).

On the other hand in Theorem 5.6.21 together with (5.6.19) and (5.6.20) we obtain the implicates (prime implicates) of f and so we may ask for the implicate counterpart of Theorem 5.5.8 too.

Thus now we show how to determine directly all representations (reduced representations of f by implicates (prime implicates).

5.7.7 Definition

Let $\mathcal{F}(f)$ $(\mathcal{F}_p(f))$ be the set of all implicates (prime implicates).

We define

$$\begin{aligned}
\mathcal{I}^*(f) &:= \{B \subseteq \mathcal{F}(f) : \min_{I^* \in B} I^* = f\}, \\
\mathcal{I}_p^*(f) &:= \{B \subseteq \mathcal{F}_p(f) : \min_{I^* \in B} I^* = f\}, \\
\mathcal{R}^*(f) &:= \{B \in \mathcal{I}^*(f) : \min_{I^* \in K} \neq f \text{ for } K \subset B\} = R(\mathcal{I}^*(f)) \\
\mathcal{R}_p^*(f) &:= \{B \in \mathcal{I}_p^*(f) : \min_{I^* \in K} \neq f \text{ for } K \subset B\} = R(\mathcal{I}_p^*(f)) \\
&= \mathcal{R}^*(f) \cap \mathcal{I}_p^*(f).
\end{aligned}$$

Obviously these four sets uniquely determine the sets of all representations (reduced representations) of f by implicates (prime implicates), and Definition 5.7.7 is the implicate version of Definition 5.5.4. Again $\mathcal{R}_p^*(f)$ is not empty, and the relations between the sets correspond to the relations given by the figure to Definition 5.5.4 in the implicant case.

5.7.8 Proposition

Let $B \subset \mathcal{F}(f)$ $(\mathcal{F}_p(f))$. Then $B \in \mathcal{I}^*(f)$ $(\mathcal{I}_p^*(f))$ if and only if for each $x \in \{f < y_k\}$, there is at least one $I^* \in B$ with $I^*(x) = f(x)$.

This follows from Definition 5.6.1 and Definition 5.7.7. The next statement is evident .

5.7.9 Proposition

$$f = \min_{I^* \in \mathcal{F}(f)} I^* = \min_{I^* \in \mathcal{F}_p(f)} I^*.$$

Thus the function f equals the minimum of all its implicates (prime implicates).

Now we state the announced counterpart to Theorem 5.5.8.

5.7.10 Theorem

For $i \in \{0, ..., k-1\}$ let $\{f = y_i\} = \{\alpha_{i1}, ..., \alpha_{iq_i}\}$.
Define for $i \in N_k, \kappa \in N_{q_i}$:

$$\mathcal{M}^*_{i\kappa}(f) := \{I^* \in \mathcal{F}_{y_i}(f) : I^*(\alpha_{i\kappa}) = y_i\},$$

$$(\mathcal{F}_{y_i} \text{ defined by } (5.6.17)),$$

$$\mathcal{M}^*_{i\kappa p}(f) := \{I^* \in \mathcal{F}_{y_ip}(f) : I^*(\alpha_{i_\kappa}) = y_i\}$$

$$(\mathcal{F}_{y_ip} \text{ defined by } (5.6.18)),$$

$$\mathcal{I}^*_i(f) := \mathcal{S}(\mathcal{M}^*_{i1}(f), ..., \mathcal{M}^*_{iq_i}(f)),$$

$$\mathcal{I}^*_{ip}(f) := \mathcal{S}(\mathcal{M}^*_{i1p}(f), ..., \mathcal{M}_{iq_ip}(f))$$

($\mathcal{S}$ defined by 2.4.13, see also Theorem 3.4.14).

Then

(5.7.11) $\mathcal{I}^*(f) = \{\dot{\bigcup}_{i=0}^{k-1} B_i : B_i \in \mathcal{I}^*_i(f) \text{ for } i = 0, ..., k-1\},$

(5.7.12) $\mathcal{I}^*_p(f) = \{\dot{\bigcup}_{i=1}^{k-1} B_i : B_i \in \mathcal{I}^*_{ipl}(f) \text{ for } i = 0, ..., k-1\},$

(5.7.13) $\mathcal{R}^*(f) = \{\dot{\bigcup}_{i=0}^{k-1} B_i : B_i \in R(\mathcal{I}^*_i(f)) \text{ for } i = 0, ..., k-1\},$

(5.7.14) $\mathcal{R}^*_p(f) = \{\dot{\bigcup}_{i=0}^{k-1} B_i : B_i \in R(\mathcal{I}^*_{ip}(f)) \text{ for } i = 0, ..., k-1\}.$

Remark

$\mathcal{M}^*_{i\kappa}(f) \neq \emptyset$, $\mathcal{M}^*_{i\kappa p}(f) \neq \emptyset$ for $\kappa \in N_{q_i}$,
$\bigcup_{\kappa \in N_{q_i}} \mathcal{M}^*_{i\kappa}(f) = \mathcal{F}_{y_i}(f)$, $\bigcup_{\kappa \in N_{q_i}} \mathcal{M}^*_{i\kappa p}(f) = \mathcal{F}_{y_ip}(f)$, for $i \in N_k$.

We omit the proof of Theorem 5.7.10 since it is similar to 5.5.8.

In the implicate case we may also give a corresponding supplement to the last result.

5.7.10' Supplement

For $i \in \{0, ..., k-1\}$, let

$$\mathcal{F}_{y_i}(f) = \{I^*_{i1}, ..., I^*_{it_i}\},$$
$$\mathcal{F}_{y_i p}(f) = \{I^{*'}_{i1}, ..., I^{*'}_{it'_i}\},$$

further $\mathcal{I}^*_i(f)$ and $\mathcal{I}^*_{ip}(f)$ as in Theorem 5.7.10.
For $N \subseteq N_{t_i}$, define

$$M^*_N := \left\{ x \in \{f = y_i\} : I^*_{i\tau}(x) = \begin{cases} y_i & \text{for } \tau \in N \\ y_k & \text{otherwise} \end{cases} \right\}$$

and for $N \subseteq N_{t'_i}$

$$M^{*'}_N := \left\{ x \in \{f = y_i\} : I^{*'}_{i\tau}(x) = \begin{cases} y_i & \text{for } \tau \in N \\ y_k & \text{otherwise} \end{cases} \right\}.$$

Let
$\{N \subseteq N_{t_i} : M^*_N \neq \emptyset\} = \{N^*_{i1}, ..., N^*_{iu_i}\}$,
$\{N \subseteq N_{t'_i} : M^{*'}_N \neq \emptyset\} = \{N^{*'}_{i1}, ..., +N^{*'}_{iu'_i}\}$.
For $j \in N_{u_i}$, define
$\mathcal{N}^*_{ij}(f) := \{I^*_{i\tau} \in \mathcal{F}_{y_i}(f) : \tau \in N^*_{ij}\} = \{I^* \in \mathcal{F}_{y_i}(f) : I^*(x) = y_i \text{ for } x \in M^*_{N^*_{ij}}\}$
and for $j \in N_{u'_i}$
$\mathcal{N}^*_{ijp}(f) := \{I^*_{i\tau} \in \{_{y_i p}(f) : \tau \in N^{*'}_{ij}\} = \{I^* \in \mathcal{F}_{y_i p}(f) : I^*(x) = y_i \text{ for } x \in M^*_{N^{*'}_{ij}}\}$.
Then, for $i \in \{0, ..., k-1\}$, we have

$$\{\mathcal{N}^*_{i1}(f), ..., \mathcal{N}^*_{iu_i}(f)\} = \bigcup_{j \in N_{q_i}} \{\mathcal{M}^*_{ij}(f)\},$$
$$\{\mathcal{N}^*_{i1p}(f), ..., \mathcal{N}^*_{iu'_i}(f)\} = \bigcup_{j \in N_{q_i}} \{\mathcal{M}_{ijp}(f)\}$$

and

$$\mathcal{I}^*_i(f) = \mathcal{S}(\mathcal{N}^*_{i1}(f), ..., \mathcal{N}^*_{iu_i}(f)),$$
$$\mathcal{I}^*_{ip}(f) = \mathcal{S}(\mathcal{N}^*_{i1p}(f), ..., \mathcal{N}^*_{iu'_i p}(f)).$$

Remark

$\{\mathcal{N}^*_{i1}(f), ..., \mathcal{N}^*_{iu_i}(f)\}$ is the set of all **different** elements of the sequence $\mathcal{M}^*_{i1}(f), ..., \mathcal{M}^*_{iq_i}(f)$.
With the notation

$$I^*_{ij} = y_i + \overline{y}_i F_{ij} \text{ for } j \in N_{t_i} \quad (\text{i.e } I^*_{ij} \in \mathcal{F}_{y_i}(f))$$

we obtain

$$M^*_N = \Big\{x \in \{f = y_i\} : \prod_{j\in N}(1 - F_{ij}) \prod_{j\in N_{t_i}\setminus N} F_{ij} = 1\Big\}.$$

With $F_{ij} = 1_{\overline{Q_{ij}}} = 1 - 1_{Q_{ij}}$ (Q_{ij} a cube) we obtain

$$M^*_N = \{f = y_i\} \cap \bigcap_{j\in N} Q_{ij} \cap \bigcap_{j\in N_{t_i}\setminus N} \overline{Q_{ij}}.$$

The corresponding holds in case of prime implicates.

Chapter 6

Applications

6.1 Reliability Structure of Technical Systems

In reliability theory we consider for example technical systems which consist of n subsystems. Usually the reliability behaviour of the system is described by an isotone Boolean function (so-called system function) $S(x_1, ..., x_n)$ of the n Boolean variables $x_1, ..., x_n$. The variable x_i is coordinated to the state of the i-th subsystem by

$$x_i := \begin{cases} 1 & \text{if the } i\text{-th subsystem works} \\ 0 & \text{otherwise .} \end{cases}$$

The system behaviour is described by

$$S(x_1, ..., x_n) := \begin{cases} 1 & \text{if the system works} \\ 0 & \text{otherwise .} \end{cases}$$

Only the states "working" and "not working" are distinguished.

Generalizing this model we assume that the system as well as the subsystems may take more than two states. A suitable description of such a system is obviously given by a discrete function

$$f : \times \{0, ..., k_i\} \to \{0, ..., k\},$$

where $f(x_1, ..., x_n)$ is the state of the system, if the i-th subsystem is in the state x_i respectively.

Now then the i-th subsystem takes the states described by $x_i = 0, ..., k_i$ $(i \in N_n)$ and the system takes the states described by $f(x_1, ..., x_n) = 0, ..., k$. In the most case of practical relevance we suppose that $f(x_1, ..., x_n)$ is isotone. Then all results of Section 5.2 may be used to give representations of f by implicants or the representation of the indicators $1_{\{f \geq y\}}$ by implicants. In Section 5.3 also the implicants of the indicators $1_{\{f=y\}}$ are treated.

Example

The function $f(x_1, x_2)$ considered in the example of Section 5.3 is a simple system function of a system with two subsystems. System and subsystems take five states.

6.2 Classification (Valuation) of Objects

6.2.1 Classification by Discrete Functions

The general classification problem can be stated in the following way.

Let S be a set of any objects, e.g. persons, animals, plants, technical products, commodities etc.

We wish to classify the set S into a suitable number of classes in such a way that all objects belonging to the same class may be regarded as equivalent in some sense, e.g. with respect to the business of a person or use of a thing.

We assume that each object is characterized by n different attributes where each attribute has a fixed number of possible realisations.

Consider as an example a (passenger) car. Then some important attributes are the power, the consumption of petrol, the security, the price, etc.

Except of the security these attributes may be characterized by numbers, respectively by intervals of numbers (power in kilowatt, consumption in litres per hundred kilometres, price in marks or dollars).

Clearly such a classification need not be a real set S of objects. We also may consider them as a scheme allowing a suitable judgement of objects e.g with the aid of a questionaire.

Now we propose the following classification procedure by using binary functions respectively discrete functions.

Each object is characterized by n attributes. The $\imath$–th attribute ($\imath \in N_n$) is valued by one of the numbers $0, 1, ..., k_i$ (or more generally by $a_{i0}, ..., a_{ik_i}$ with $a_{i0} < ... < a_{ik_i}$). The valuation of all n attributes then corresponds to exactly one point

$$\jmath := (\jmath_1, ..., \jmath_n) \in K := \bigtimes_{i=1}^{n} \{0, ..., k_i\}.$$

If we wish a classification into $k+1$ different classes (of equivalent objects), we have to divide K into $k+1$ mutually disjoint nonempty subsets $\Gamma_0, ..., \Gamma_k$. This division (partition) of K is arbitrary from a mathematical viewpoint. In reality it needs an expert to find a reasonable partition.

An object with the valuation corresponding to $\jmath := (\jmath_1, ..., \jmath_n)$ now belongs to the class κ corresponding to Γ_κ ($\kappa \in \{0, ..., k\}$) if and only if $\jmath \in \Gamma_\kappa$.

Now it is obvious to introduce variables $x_1, ..., x_n$ taking the values $x_i = 0, ..., k_i$ for $\imath \in N_n$ and the discrete function $f(x_1, ..., x_n)$ taking the values $0, ..., k$ in such a way that $f(x_1, ..., x_n) = \kappa$ for $(x_1, ..., x_n) \in \Gamma_\kappa, \kappa \in \{0, ..., k\}$. We interprete this as follows:

"The variable x_i ($i \in N_n$) takes the value $j_i \in \{0, ..., k_i\}$ if and only if the i–th attribute of the considered object is valued by the number j_i. An object belongs to the class $\kappa \in \{0, ..., k\}$ if and only if $(x_1, ..., x_n) \in \Gamma_\kappa$, i.e. if and only if $f(x_1, ..., x_n) = \kappa$."

Obviously $f(x_1, ..., x_n) = \kappa$ is the same as $1_{\Gamma_\kappa}(x_1, ..., x_n) = 1$.

We remark that the valuation of the i–th attribute by the numbers $0, ..., k_i$ does not imply that an object with the value j_i is "better" than an object with a value $j_i' < j_i$. Also any object out of the class Γ_κ need not be "better" than an object out of $\Gamma_{\kappa'}$ with $\kappa' < \kappa$. However in many cases (e.g. school marks) such a quantitative valuation is usual.

6.2.2 Characterization of Classes by Implicants

From Chapter 2 we know that each indicator $1_{\Gamma_\kappa} = 1_{\Gamma_\kappa}(x_1, ..., x_n)$ has representations by implicants, especially minimal representations by (prime) implicants. This corresponds to representations of the set Γ_Γ as a union of cubes. Such a representation is "optimal" in so far as it contains a minimal number of cubes.

Let us assume that

$$1_{\Gamma_\kappa} = \max_{\rho \in N_r} C_\rho$$

is a minimal representation by the (prime) implicants $C_1, ..., C_r$. Each C_ρ is a cube indicator to a cube $\times P_{\rho i} \subset K$. The optimal representation of Γ_κ as union of cubes is given by

$$\Gamma_\kappa = \bigcup_{\rho \in N_r} \times P_{\rho i}.$$

Now we may describe the class κ as follows:

> "An object belongs to the class κ if and only if for at least one $\rho \in N_r$ it holds
>
> $$(x_1, ..., x_n) \in \times P_{\rho i},$$
>
> i.e. $x_1 \in P_{\rho 1}$ and $x_2 \in P_{\rho 2}$ and ... and $x_n \in P_{\rho n}$."

6.2.3 Characterization of Classes by Implicates

According to Chapter 3 we have minimal representations

$$1_{\Gamma_\kappa} = \min_{\sigma \in N_s} F_\sigma$$

by (prime) implicates. Each F_σ is an anticube indicator to an anticube $\bigcup Q^*_{\sigma i} \subset K$ (see (2.1.4)). The corresponding "optimal" representation of Γ_κ as an intersection of anticubes is given by

$$\Gamma_\kappa = \bigcap_{\sigma \in N_s} \bigcup Q^*_{\sigma i}.$$

We obtain a further description of the class κ as follows:

"An object belongs to the class κ if and only if for each $\sigma \in N_s$ it holds

$$(x_1, ..., x_n) \in \bigcup Q^*_{\sigma i},$$

i.e. (see (2.1.4))

$$x_1 \in Q_{\sigma 1} \quad \text{or} \quad x_2 \in Q_{\sigma 2} \quad \text{or} \; x_n \in Q_{\sigma n}."$$

Example

The discrete function $f(x_1, x_2)$ described in the example at the end of Section 5.3 may be used to give a simple classification, where the objects are valuated by only two attributes each of them taking the values $0, ..., 4$. Each object belongs to one of the classes $0, ..., 4$.

Chapter 7

A Class of Finite Boolean Algebras

In this chapter we will show that the set of all binary functions defined on a finite Cartesian Product M may be considered as a special case of a more general class of so–called Boolean algebras e.g. systems of events in probability theory or truth function systems used in propositional logic. This is of practical importance because implicants (prime implicants) and implicates (prime implicates) may be also defined in such Boolean algebras. Moreover it will appear that all results concerning implicants and implicates of binary functions may be translated in a simple way into the general model of a certain type of a Boolean algebra.

To clarify the problem we may first consider a set algebra in a set Ω. We ask for the smallest algebra containing some subsetes $A_1, ..., A_n$ of Ω. It is easy to show that this algebra contains all 2^n intersections of the type $\overset{(-)}{A_1} \cap ... \cap \overset{(-)}{A_n}$ (containing always A_i or the complement $\overline{A}_i$ of A_i), futher all possible unions of these intersections.

With each A_i – we may suppose $\emptyset \neq A_i \neq \Omega$ – also the complement $\overline{A}_i$ belongs to our algebra. We may regard $\{A_1, \overline{A}_1, ..., A_n, \overline{A}_n\}$ as its generating system. Obviously a set A_i and its complement $\overline{A}_i$, both regarded as events form an alternative: either A_i occurs or $\overline{A}_i$ occurs (i.e. A_i does not occur). It seems to be obvious to generalize such alternatives to partitions of Ω with more than two sets.

In a partition $\{A_1, ..., A_m\}$ of Ω exactly one of the events $A_1, ..., .A_m$ occurs (i.e. $\bigcup_{i \in N_m} A_i = \Omega, A_i \neq \emptyset$ for $i \in N_m$ and $A_i \cap A_j = \emptyset$ for $i \neq j$).

Such partitions often occur in the real world. To discribe the season of a year e.g. it seems better to use the four statements (the partition)

"it is spring" ,

"it is summer" ,

"it is autumn" ,

"it is winter" ,

instead of the alternative

"it is summer" , "it is not summer".

In a similar way we may interprete the truth functions of m propositions $p_1, ..., p_m$ as a partition if exactly one of them is true. Then $p_1 \vee ... \vee p_m$ is true and $p_i \wedge p_j$

is false for $i \neq j$, i.e. exactly one of the propositions is true (and often we do not know which of them is true; principially each of them may be true).

Now let $\{A_{10}, .., A_{1k_1}\}, ..., \{A_{n0}, ..., A_{nk_n}\}$ be n partitions of Ω.

We ask for the smallest algebra in Ω containing $\{A_{10}, ..., A_{nk_n}\}$. We call it the algebra in Ω generated by the n given partitions.

Now more generally we will consider Boolean algebras, especially Boolean algebras generated by partitions.

Then we shall apply the results to special Boolean algebras as set algebras, indicator algebra, algebras of classes of propositions, truth function algebras. The indicators 1_Γ treated in Chapter 2 and Chapter 3 will appear as elements of such an indicator algebra. Moreover we shall show, that the implicants and implicates of elements of a Boolean algebra generated by partitions may be obtained from the implicants and implicates of indicators.

7.1 Boolean Algebras

First we need some definitions.

7.1.1 Definition

An *order relation* $\leq$ on a set $\mathcal{B}$ is a relation with the following properties:
For all $x, y, z \in \mathcal{B}$ holds

$$x \leq x$$
$$x \leq y,\ y \leq z \text{ implies } x \leq z\ ,$$
$$x \leq y,\ y \leq x \text{ implies } x = y\ .$$

7.1.2 Definition

Let $\mathcal{B}$ be a set with an order relation $\leq$.
If $x, y, z \in \mathcal{B}$ and

$$x \leq z, y \leq z \text{ and } z \leq z' \quad \text{for each } z' \in \mathcal{B} \text{ with } x \leq z', y \leq z'$$

then we call z the *supremum* of x and y, written $z = x \vee y$.
If $x, y, z \in \mathcal{B}$ and

$$z \leq x, z \leq y \text{ and } z' \leq z \quad \text{for each } z' \in \mathcal{B} \text{ with } z' \leq x, z' \leq y$$

then we call z the *infimum* of x and y, written $z = x \wedge y$.

Obviously it holds $x \vee y = y \vee x$, $x \wedge y = y \wedge x$ (so-called commutative law). Now we may define a Boolean algebra.

7.1.3 Definition

A *Boolean algebra* is a set $\mathcal{B}$ with an order relation $\leq$ such that

(a) $x \vee y$ and $x \wedge y$ exists for all $x, y \in \mathcal{B}$;

(b) $x \wedge (y \vee z) = (x \wedge y) \vee (x \wedge z)$ and
$x \vee (y \wedge z) = (x \vee y) \wedge (x \vee z)$
holds for all $x, y, z \in \mathcal{B}$ (so-called distributive laws);

(c) there exists a least element (*0-element*) $\mathbb{O} \in \mathcal{B}$ and a greatest element (*1-element*) $\mathbb{1} \in \mathcal{B}$ satisfying $\mathbb{O} \leq x \leq \mathbb{1}$ for all $x \in \mathcal{B}$;

(d) for all $x \in \mathcal{B}$ there exists a (unique) element $\neg x \in \mathcal{B}$ (*complement of* x) satisfying

$$x \vee \neg x = \mathbb{1}, \; x \wedge \neg x = \mathbb{O}.$$

Remark

A set algebra in a set Ω, i.e. a system $\mathcal{A}$ of subsets of Ω satisfying

(a) $\Omega \in \mathcal{A}$,
(b) $A \in \mathcal{A}, A' \in \mathcal{A}$ implies $A \cup A' \in \mathcal{A}$,
(c) $A \in \mathcal{A}$ implies $\overline{A} \in \mathcal{A}$

is always a Boolean algebra with the 0-element $\emptyset$, 1-element Ω, $A \vee A' := A \cup A', A \wedge A' := A \cap A', \neg A := \overline{A}$ and the order relation $A \leq A' :\Leftrightarrow A \subseteq A'$.

This follows from Theorem 7.1.5.

Next we state a lemma used in the further investigations.

7.1.4 Lemma

Let Ω be a set, $\mathcal{P}(\Omega)$ the set of all subsets of Ω (power set of Ω), $\mathcal{C} \subseteq \mathcal{P}(\Omega)$ and $\{B_A : A \in \mathcal{C}\}$ a set with a relation $\leq$ satisfying $B_A \leq B_{A'} \Leftrightarrow A \subseteq A'$. Then $\leq$ is an order relation on $\{B_A : A \in \mathcal{C}\}$.

Proof

$A \subseteq A$ yields $B_A \leq B_A$. From $B_A \leq B_{A'}, B_{A'} \leq B_{A''}$ we obtain $A \subseteq A', A' \subseteq A''$, thus $A \subseteq A''$ and so $B_A \leq B_{A''}$. From $B_A \leq B_{A'}$ and $B_{A'} \leq B_A$ we obtain

$A \subseteq A'$ and $A' \subseteq A$, thus $A = A'$ and so $B_A = B_{A'}$. □

The following theorem is the basis of the construction of several Boolean algebras connected with a set algebra in a set Ω.

7.1.5 Theorem

Let Ω be a set, $\mathcal{A} \subseteq \mathcal{P}(\Omega)$ and $\mathcal{B} := \{B_A : A \in \mathcal{A}\}$ a set with the order relation $\leq$ according to 7.1.4.

Then the following are equivalent:

(a) $\mathcal{A}$ is a set algebra in Ω.

(b) $\mathcal{B}$ is a Boolean algebra with the order relation $\leq$ and

$$\mathbb{O} = B_\phi, \mathbb{1} = B_\Omega, B_A \vee B_{A'} = B_{A\cup A'}, B_A \wedge B_{A'} = B_{A\cap A'}, \neg B_A = B_{\overline{A}} \,.$$

Proof

Suppose (a) and $A, A', A'' \in \mathcal{A}$. From $\phi \subseteq A \subseteq \Omega$ it follows $B_\phi \leq B_A \leq B_\Omega$. Thus B_ϕ is the 0–element $\mathbb{O}$ and B_Ω is the 1–element $\mathbb{1}$. From $A, A' \subseteq A \cup A' \subseteq A''$ for each A'' with $A, A' \subseteq A''$ we obtain $B_A, B_{A'} \leq B_{A\cup A'} \leq B_{A''}$ for each $B_{A''}$ with $B_A, B_{A'} \leq B_{A''}$. Thus $B_A \vee B_{A'} = B_{A\cup A'}$. From $A'' \subseteq A \cap A' \subseteq A, A'$ for each A'' with $A'' \subseteq A, A'$ in the same way it follows $B_A \wedge B_{A'} = B_{A\cap A'}$.
Now

$$\begin{aligned} B_A \wedge (B_{A'} \vee B_{A''}) &= B_A \wedge B_{A'\cup A''} = B_{A\cap(A'\cup A'')} \\ &= B_{(A\cap A')\cup(A\cap A'')} = B_{A\cap A'} \vee B_{A\cap A''} \\ &= (B_A \wedge B_{A'}) \vee (B_A \wedge B_{A''}) \end{aligned}$$

and

$$\begin{aligned} B_A \vee (B_{A'} \wedge B_{A''}) &= B_A \vee B_{A'\cap A''} = B_{A\cup(A'\cap A'')} \\ &= B_{(A\cup A')\cap(A\cup A'')} = B_{A\cup A'} \wedge B_{A\cup A''} \\ &= (B_A \vee B_{A'}) \wedge (B_A \vee B_{A''}) \,. \end{aligned}$$

Thus the distributive laws hold.

Finally $B_A \vee B_{\overline{A}} = B_{A\cup\overline{A}} = B_\Omega = \mathbb{1}$ and $B_A \wedge B_{\overline{A}} = B_{A\cap\overline{A}} = B_\phi = \mathbb{O}$ implies $\neg B_A = B_{\overline{A}}$. Therefore (b) holds.

Now suppose (b). From $\mathbb{1} = B_\Omega \in \mathcal{B}$ it follows $\Omega \in \mathcal{A}$. Let $A, A' \in \mathcal{A}$ and so $B_A, B_{A'} \in \mathcal{B}$. Then $B_{A\cup A'} = B_A \vee B_{A'} \in \mathcal{B}$ and so $A \cup A' \in \mathcal{A}$, further

$B_{\overline{A}} = \neg B_A \in \mathcal{B}$ and so $\overline{A} \in \mathcal{A}$. Therefore (a) holds. □

7.2 Boolean Algebras Generated by Finite Partitions

Assume $\mathcal{B}$ to be a Boolean algebra and $\mathcal{C}$ any nonempty subsystem of $\mathcal{B}$. Then we may ask for the smallest Boolean subalgebra of $\mathcal{B}$ containing $\mathcal{C}$.

The following theorem states the existence and uniqueness of such a Boolean subalgebra.

7.2.1 Theorem

Let $\mathcal{B}$ be a Boolean algebra and $\emptyset \neq \mathcal{C} \subseteq \mathcal{B}$. Then there exists a unique Boolean algebra $\mathcal{B}(\mathcal{C}) \subseteq \mathcal{B}$ such that

(a) $\mathcal{C} \subseteq \mathcal{B}(\mathcal{C})$,

(b) $\mathcal{B}(\mathcal{C}) \subseteq \mathcal{B}'$ for each Boolean algebra $\mathcal{B}'$ with $\mathcal{C} \subseteq \mathcal{B}' \subseteq \mathcal{B}$.

The Boolean algebra $\mathcal{B}(\mathcal{C})$ contains exactly all elements of $\mathcal{C}$ as well as all further elements of $\mathcal{B}$ which may be obtained stepwise from the elements of $\mathcal{C}$ by the Boolean operations $\vee, \wedge, \neg$.

We call $\mathcal{B}(\mathcal{C})$ the Boolean algebra generated by $\mathcal{C}$. We remark that $\mathcal{B}(\mathcal{C})$ equals $\mathcal{C}$ if an only if $\mathcal{C}$ is a Boolean algebra.

Proof

Let $\mathcal{B}(\mathcal{C})$ be given as stated above. Obviously $\mathcal{B}(\mathcal{C})$ is a Boolean algebra. Now let $\mathcal{B}' \subseteq \mathcal{B}$ be a Boolean algebra with $\mathcal{C} \subseteq \mathcal{B}'$. Then by construction of $\mathcal{B}(\mathcal{C})$ also $\mathcal{B}(\mathcal{C}) \subseteq \mathcal{B}'$. Thus $\mathcal{B}(\mathcal{C})$ satisfies (a) and (b). The uniqueness is evident. □

Now we ask especially for Boolean algebras which are generated by a set of partitions of the 1–element of a given Boolean algebra, since such Boolean algebras have a close relation to the indicators (binary functions) introduced in Chapter 1.

First we define partitions.

7.2.2 Definition

Let $\mathcal{B}$ be a Boolean algebra with 0–element $\mathbb{O}$ and 1–element $\mathbb{1}$. Then $\{B_1, ..., B_r\} \subset \mathcal{B}$ with $r > 1$ is called a *partition* of $\mathbb{1}$ if

$$B_i \neq \mathbb{O}, B_i \wedge B_j = \mathbb{O} \text{ for } i, j \in N_r, i \neq j \text{ and } \bigvee_{i \in N_r} B_i := B_1 \vee ... \vee B_r = \mathbb{1}.$$

The next theorem yields an explicite representation of a Boolean algebra generated by n partitions of $\mathbb{1}$.

7.2.3 Theorem

Let $\mathcal{B}$ be a Boolean algebra and

$$\{B_{10},...,B_{1k_1}\},...,\{B_{n0},...,B_{nk_n}\}$$

n partitions of $\mathbb{1}$.

Define

$$K := \bigtimes_{i\in N_n} \{0,...,k_i\} ,$$

$$B_j := \bigwedge_{i\in N_n} B_{ij_i} := B_{1j_1} \wedge ... \wedge B_{nj_n} \quad \text{for each } j := (j_1,...,j_n) \in K ,$$

$$B_\Gamma := \bigvee_{j\in\Gamma} B_j \qquad \text{for each } \Gamma \subseteq K \text{ , } B_\phi := \mathbb{O} .$$

Then we have

$$\mathcal{B}(\{B_{10},...,B_{nk_n}\}) = \{B_\Gamma : \Gamma \subseteq K\}.$$

For the elements B_Γ of $\mathcal{B}(\{B_{10},...,B_{nk_n}\})$ we have

a) $B_\phi = \mathbb{O}$,
b) $B_K = \mathbb{1}$,
c) $B_\Gamma \vee B_{\Gamma'} = B_{\Gamma\cup\Gamma'}$,
d) $B_\Gamma \wedge B_{\Gamma'} = B_{\Gamma\cap\Gamma'}$,
e) $\neg B_\Gamma = B_{\overline{\Gamma}}$ with $\overline{\Gamma} := K \setminus \Gamma$.

The elements $B_j, j \in K$ with $B_j \neq \mathbb{O}$ are the *atoms* of $\mathcal{B}(\{B_{10},...,B_{nk_n}\})$, i.e. for each B_j it holds

$$B_j \wedge B = \begin{cases} \mathbb{O} \\ \text{or} \\ B_j \end{cases} \quad \text{for } B \in \mathcal{B}(\{B_{10},...,B_{nk_n}\}) .$$

The atoms form a partition of $\mathbb{1}$.

Proof

a) By definition we have $B_\phi = \mathbb{O}$.

b) $$B_K = \bigvee_{j\in K} B_j = \bigvee_{(j_1,...,j_n)\in K} \bigwedge_{i=1}^{n} B_{ij_i} = \bigwedge_{i=1}^{n} \bigvee_{k=0}^{k_i} B_{ik} = \bigwedge_{i=1}^{n} \mathbb{1} = \mathbb{1} .$$

c)
$$B_\Gamma \vee B_{\Gamma'} = \bigvee_{j\in\Gamma} B_j \vee \bigvee_{j\in\Gamma'} B_j = \bigvee_{j\in\Gamma} B_j \vee \bigvee_{j\in\Gamma\cap\Gamma'} B_j \vee \bigvee_{j\in\Gamma'\setminus\Gamma} B_j$$
$$= \bigvee_{j\in\Gamma} B_j \vee \bigvee_{j\in\Gamma'\setminus\Gamma} B_j = \bigvee_{j\in\Gamma\cup\Gamma'} B_j = B_{\Gamma\cup\Gamma'} .$$

d) The partition property $B_{i\kappa} \wedge B_{i\kappa'} = \mathbb{O}$ for $i = 1, ..., n; \kappa, \kappa' \in \{0, ..., k_i\}$, $\kappa \neq \kappa'$ implies $B_j \wedge B_k = \mathbb{O}$ for $j, k \in K, j \neq k$. Now

$$B_\Gamma \wedge B_{\Gamma'} = (\bigvee_{j\in\Gamma\cap\Gamma'} B_j \vee \bigvee_{j\in\Gamma\setminus\Gamma'} B_j) \wedge (\bigvee_{j\in\Gamma\cap\Gamma'} B_j \vee \bigvee_{j\in\Gamma'\setminus\Gamma} B_j)$$
$$= \bigvee_{j\in\Gamma\cap\Gamma'} B_j \vee (\bigvee_{\Gamma\cap\Gamma'} B_j \wedge \bigvee_{j\in\Gamma'\setminus\Gamma} B_j) \vee (\bigvee_{j\in\Gamma|\Gamma'} B_j) \wedge (\bigvee_{j\in\Gamma|\Gamma'} B_j)$$
$$\vee (\bigvee_{j\in\Gamma\setminus\Gamma'} B_j \wedge \bigvee_{j\in\Gamma'\setminus\Gamma} B_j)$$
$$= \bigvee_{j\in\Gamma\cap\Gamma'} B_j \vee \mathbb{O} \vee \mathbb{O} \vee \mathbb{O} = \bigvee_{j\in\Gamma\cap\Gamma'} B_j = B_{\Gamma\cap\Gamma'} .$$

e)
$$B_\Gamma \wedge B_{K\setminus\Gamma} = \bigvee_{j\in\Gamma} B_j \wedge \bigvee_{j\in\overline{\Gamma}} B_j = \mathbb{O} \quad \text{(see d))} ,$$
$$B_\Gamma \vee B_{\overline{\Gamma}} = B_{\Gamma\cup\overline{\Gamma}} = B_K = \mathbb{1} \quad \text{(see c) and a))} .$$

Now a),...,e) imply that $\{B_\Gamma : \Gamma \subseteq K\}$ is a Boolean algebra (from $\{B_\Gamma : \Gamma \subseteq K\} \subseteq \mathcal{B}$ it follows that $\{B_\Gamma : \Gamma \subseteq K\}$ is distributive).

Next we show that $B_{i_oj_o}$ for each $i_0 \in N_n, j_0 \in \{0, ..., k_{i_0}\}$ belongs to $\{B_\Gamma : \Gamma \subseteq K\}$ implying $\{B_{10}, ..., B_{nk_n}\} \subseteq \{B_\Gamma : \Gamma \subseteq K\}$. Since $\bigvee_{j_i=0}^{k_i} B_{ij_i} = \mathbb{1}$ for each $i \in N_n$ we have

$$B_{i_0j_0} = B_{i_0j_0} \wedge (\bigwedge_{i\in N_n\setminus\{i_0\}} \bigvee_{j_i=0}^{k_i} B_{ij_i}) = B_{i_0j_0} \wedge (\bigvee_{\substack{j_i\in\{0,...,k_i\}\\ \text{for } i\in N_n\setminus\{i_0\}}} \bigwedge_{i\in N_n\setminus\{i_0\}} B_{ij_i})$$
$$= \bigvee_{\substack{j_{i_0}=j_0\\ j_i\in\{0,...,k_i\}\\ \text{for } i\in N_n\setminus\{i_0\}}} \bigwedge_{i\in N_n} B_{ij_i} = B_{\Gamma'}$$

with $\Gamma' = \{(j_1, ..., j_n) \in K : j_{i_0} = j_0\}$. This implies $B_{i_0j_0} \in \{B_\Gamma : \Gamma \subseteq K\}$.

Now suppose $\mathcal{B}_0$ to be a Boolean algebra with $\{B_{10}, ..., B_{nk_n}\} \subseteq \mathcal{B}_0 \subseteq \mathcal{B}$. Then evidently $\{B_\Gamma : \Gamma \subseteq K\} \subseteq \mathcal{B}$. Thus $\{B_\Gamma : \Gamma \subseteq K\}$ is the Boolean algebra generated by $\{B_{10},...,B_{nk_n}\}$. Further $B_j \wedge B_k = \emptyset$ for $j \neq k$ and $\bigvee_{j\in K} B_j = \mathbb{1}$ (see Proof of d) and b)). Finally it holds

$$B_j \wedge B_\Gamma = \begin{cases} \emptyset & \text{if } j \notin \Gamma \\ B_j & \text{if } j \in \Gamma \,. \end{cases}$$

□

Remark

The following are equivalent:

(a) $B_j \wedge B = \begin{cases} \emptyset \\ \text{or} \\ B_j \end{cases}$ for $B \in \mathcal{B}(\{B_{10},...,B_{n,k_n}\})$

(b) $B \in \mathcal{B}(\{B_{10},...,B_{nk_n}\}), B \leq B_j$ implies $B = \begin{cases} \emptyset \\ \text{or} \\ B_j \end{cases}$.

Proof

Suppose (a) and $B \in \mathcal{B}(\{B_{10},...,B_{nk_n}\}), B \leq B_j$. Then $B = B_j \wedge B$ and so $B = \begin{cases} \emptyset \\ \text{or} \\ B_j \end{cases}$.

Suppose (b) and $B \in \mathcal{B}(\{B_{10},...,B_{nk_n}\})$. Then $B_j \wedge B \in \mathcal{B}(\{B_{10},...,B_{nk_n}\})$ and $B_j \wedge B \leq B_j$ implies $B_j \wedge B = \begin{cases} \emptyset \\ \text{or} \\ B_j \end{cases}$. □

Comment

Let $\Gamma, \Gamma' \subseteq K$ and $\Gamma \subset \Gamma'$. Then Theorem 7.2.3 d) implies $B_\Gamma \wedge B_{\Gamma'} = B_\Gamma$ and so $B_\Gamma \leq B_{\Gamma'}$. But $B_\Gamma \leq B_{\Gamma'}$ does not imply $\Gamma \subseteq \Gamma'$. Suppose e.g. $\Gamma \setminus \Gamma' \neq \emptyset$ and $B_j = \emptyset$ for $j \in \Gamma \setminus \Gamma'$. Then $\Gamma \not\subseteq \Gamma'$ but $B_{\Gamma\setminus\Gamma'} = \bigvee_{j\in\Gamma\setminus\Gamma'} B_j = \emptyset$ and so $B_\Gamma = B_{\Gamma\cap\Gamma'} \vee B_{\Gamma\setminus\Gamma'} = B_{\Gamma\cap\Gamma'} \vee \emptyset = B_{\Gamma\cap\Gamma'} = B_\Gamma \wedge B_{\Gamma'}$, i.e. $B_\Gamma \leq B_{\Gamma'}$.
On the other hand from the following lemma it follows that $\Gamma \subseteq \Gamma'$ is equivalent to $B_\Gamma \leq B_{\Gamma'}$ for alle $\Gamma, \Gamma' \subseteq K$ if and only if $B_j \neq \emptyset$ for each $j \in K$.

7.2.4 Lemma

Let $K \neq \emptyset$ be a finite set and

$$\{B_\Gamma : \Gamma \subseteq K\}$$

a Boolean algebra with $B_\Gamma := \bigvee_{j\in\Gamma} B_j$ for $\Gamma \subseteq K$, $B_\phi := \mathbb{O}, B_K = \mathbb{1}$ and

$$B_j \wedge B_k = \mathbb{O} \quad \text{for } j,k \in K, j \neq k.$$

Then the following are equivalent:

a) $B_j \neq \mathbb{O}$ for each $j \in K$.
b) For all $\Gamma, \Gamma' \subseteq K$ holds:
$B_\Gamma \leq B_{\Gamma'}$ if and only if $\Gamma \subseteq \Gamma'$.

Remark

From b) it follows

c) $B_\Gamma < B_{\Gamma'}$ (i.e. $B_\Gamma \leq B_{\Gamma'}, B_\Gamma \neq B_{\Gamma'}$) if and only if $\Gamma \subset \Gamma'$
d) $B_\Gamma = B_{\Gamma'}$ if and only if $\Gamma = \Gamma'$
e) $B_\Gamma = B_\emptyset$ if and only if $\Gamma = \emptyset$.

Proof

1. Suppose a) and $\Gamma \subseteq \Gamma'$. Then (see also the proof of Theorem 7.2.3, d)) $B_\Gamma = B_{\Gamma\cap\Gamma'} = B_\Gamma \wedge B_{\Gamma'}$ and so $B_\Gamma \leq B_{\Gamma'}$.

2. Suppose a) and $\Gamma \not\subseteq \Gamma'$. Then $\Gamma \setminus \Gamma' \neq \emptyset$. This implies $B_{\Gamma\setminus\Gamma'} \neq \mathbb{O}$ since from $B_{\Gamma\setminus\Gamma'} = \mathbb{O}$ it follows $B_j = B_j \wedge B_{\Gamma\setminus\Gamma'} = \mathbb{O}$ for each $j \in \Gamma \setminus \Gamma'$ in contradiction to a).

Now assume $B_\Gamma \leq B_{\Gamma'}$. Then $B_\Gamma = B_\Gamma \wedge B_{\Gamma'} = B_{\Gamma\cap\Gamma'}$. From $B_\Gamma = B_{\Gamma\cap\Gamma'} \vee B_{\Gamma\setminus\Gamma'}$ (see proof of Theorem 7.2.3 c)) then we obtain

$$B_{\Gamma\cap\Gamma'} = B_{\Gamma\cap\Gamma'} \vee B_{\Gamma\setminus\Gamma'}$$

yielding

$$\begin{aligned}\mathbb{O} &= B_{\Gamma\cap\Gamma'} \wedge B_{\Gamma\setminus\Gamma'} = (B_{\Gamma\cap\Gamma'} \vee B_{\Gamma\setminus\Gamma'}) \wedge B_{\Gamma\setminus\Gamma'} \\ &= (B_{\Gamma\cap\Gamma'} \wedge B_{\Gamma\setminus\Gamma'}) \vee (B_{\Gamma\setminus\Gamma'} \wedge B_{\Gamma\setminus\Gamma'}) = \mathbb{O} \vee B_{\Gamma\setminus\Gamma} = B_{\Gamma\setminus\Gamma'}\end{aligned}$$

in contradiction to $B_{\Gamma\setminus\Gamma'} \neq \mathbb{O}$. Thus $B_\Gamma \not\leq B_{\Gamma'}$.

3. Suppose b). For any $j \in K$ and $j \notin \Gamma$, let $\Gamma' = \Gamma \cup \{j\}$. Then due to c) it holds $B_\Gamma < B_{\Gamma'} = B_{\Gamma\cup\{j\}} = B_\Gamma \vee B_j$. Thus $B_j \neq \mathbb{O}$, since $B_j = \mathbb{O}$ yields $B_\Gamma = B_{\Gamma'}$.□

In the following we give a survey of the properties of the Boolean algebra $\mathcal{B}(\{B_{10}, ..., B_{nk_n}\})$ stated by Theorem 7.2.3. We suppose $B_j \neq \mathbb{O}$ for all $j \in K$.

7.2.5 Properties of $\mathcal{B}(\{B_{10}, ..., B_{nk_n}\}) = \{B_\Gamma : \Gamma \subseteq K\}$

(a) The number of the atoms of $\mathcal{B}(\{B_{10}, ..., B_{nk_n}\})$ is
$|\{B_j : j \in K\}| = |K| = \prod_{i \in N_n} (k_i + 1)$,

(b) the number of the elements of $\mathcal{B}(\{B_{10}, ..., B_{nk_n}\})$ is

$|\mathcal{B}(\{B_{10}, ..., B_{nk_n}\})| = |\{B_\Gamma : \Gamma \subseteq K\}| = |\mathcal{P}(K)| = 2^{|K|} = 2^{\prod_i^n (k_i+1)}$,

(c) $B_\emptyset = \mathbb{0}$,

(d) $B_K = \mathbb{1}$,

(e) $B_j = B_{\{j\}}$ for all $j \in K$,

(f) $B_j \wedge B_k = \mathbb{0}$ for all $j, k \in K, j \neq k$,

(g) any B_j, there exists no $B \in \mathcal{B}((B_{10}, ..., B_{nk_n}\})$ with $\mathbb{0} \neq B \neq B_j$ and $B \leq B_j$ (the elements B_j are the atoms of $\mathcal{B}(\{B_{10}, ..., B_{nk_n}\})$),

(h) $B_{ij} = B_{\Gamma_{ij}}$ with $\Gamma_{ij} := \{(j_1, ..., j_k) \in K : j_i = j\}$,

(i) $B_\Gamma \vee B_{\Gamma'} = B_{\Gamma \cup \Gamma'}$,

(j) $B_\Gamma \wedge B_{\Gamma'} = B_{\Gamma \cap \Gamma'}$,

(k) $\neg B_\Gamma = B_{K \setminus \Gamma}$,

(l) $B_\Gamma \leq B_{\Gamma'}$ if and only if $\Gamma \subseteq \Gamma'$,

(m) $B_\Gamma = B_{\Gamma'}$ if and only if $\Gamma = \Gamma'$,

(n) $B_\Gamma \vee B_{\Gamma'} = \mathbb{1}$ if and only if $\Gamma \cup \Gamma' = K$,

(o) $B_\Gamma \vee B_{\Gamma'} = \mathbb{0}$ if and only if $\Gamma = \Gamma' = \emptyset$,

(p) $B_\Gamma \wedge B_{\Gamma'} = \mathbb{1}$ if and only if $\Gamma = \Gamma' = K$,

(q) $B_\Gamma \wedge B_{\Gamma'} = \mathbb{0}$ if and only if $\Gamma \cap \Gamma' = \emptyset$.

Theorem 7.1.5 and Theorem 7.2.3 enable us to establish a class of Boolean algebras which are connected with set algebras in a set Ω. Examples of them will be treated in Chapter 8. The general structure of such Boolean algebras is given by the following theorem.

7.2.6 Theorem

Let Ω be a set and $\{A_{10}, ..., A_{1k_1}\}, ..., \{A_{n0}, ..., A_{nk_n}\}$ n partitions of Ω.
Let $\mathcal{B} := \{B_A : A \subseteq \Omega\}$ be ordered by the relation $\leq$ defined in 7.1.4.
Define B_{ij} for $i \in N_n$, $j \in \{0, ..., k_i\}$ by $B_{ij} := B_{A_{ij}}$. Then $\mathcal{B}$ is a Boolean algebra with the order relation $\leq$ and

$$\mathbb{0} = B_\emptyset, \mathbb{1} = B_\Omega, B_A \vee B_{A'} = B_{A \cup A'}, B_A \wedge B_{A'} = B_{A \cap A'}, \neg B_A = B_{\overline{A}}.$$

Further

$$\{B_{10},...,B_{1k_1}\},...,\{B_{n0},...,B_{nk_n}\}$$

are partitions of $\mathbb{1}$, and the Boolean algebra generated by $\{B_{10},...,B_{nk_n}\}$ is given by

$$\mathcal{B}(\{B_{10},...,B_{nk_n}\}) = \{B_\Gamma : \Gamma \subseteq K\}$$

with $K := \bigtimes_{i \in N_n} \{0,...,k_i\}$, $B_\Gamma := \bigvee_{j \in \Gamma} B_j$ for $\Gamma \subseteq K$, $B_j := \bigwedge_{i \in N_n} B_{ij_i}$ for each $j := (j_1,...,j_n) \in K$.

Proof

The set $\mathcal{P}(\Omega)$ is an algebra in Ω. Thus $\mathcal{B} := \{B_A : A \subseteq \Omega\} = \{B_A : A \in \mathcal{P}(\Omega)\}$ is an Boolean algebra with the stated properties according to Theorem 7.1.5.

From $A_{ij} \neq \emptyset$ we obtain $B_{ij} = B_{A_{ij}} \neq \mathbb{O}$. Further $A_{ij} \cap A_{ij'} = \emptyset$ yields $B_{ij} \wedge B_{ij'} = B_{A_{ij}} \wedge B_{A_{ij'}} = B_{A_{ij} \cap A_{ij'}} = B_\emptyset = \mathbb{O}$. From $\bigcup_{j=0}^{k_i} A_{ij} = \Omega$ we obtain $\bigvee_{j=0}^{k_i} B_{ij} = B_{\bigcup_{j=0}^{k_i} A_{ij}} = B_\Omega = \mathbb{1}$.

Thus $\{B_{10},...,B_{1k_1}\},...,\{B_{n0},...,B_{nk_n}\}$ are partitions of $\mathbb{1}$. Now Theorem 7.2.3 yields the last proposition.

7.3 Representation of Boolean Elements by Implicants

In 7.3 and 7.4 we consider Boolean algebras of the type $\mathcal{B}(\{B_{10},...,B_{nk_n}\}) = \{B_\Gamma : \Gamma \subseteq K\}$ and show that the elements B_Γ have representations by Boolean implicants and Boolean implicates. Moreover we will see that all kinds of such implicants and implicates and all kinds of representations by them may be obtainded in a simple way from the corresponding quantities concerning indicators handled in Chapter 2 and 3.

First we consider representations by implicants.

7.3.1 Notation

For abbreviation we write

$$\mathcal{B}^* := \mathcal{B}(\{B_{10},...,B_{nk_n}\}) = \{B_\Gamma : \Gamma \subseteq K\}.$$

The following definitions may be regarded as generalisations of the corresponding definitions of Chapter 2.

7.3.2 Definition

Let $\emptyset \neq \mathop{\times}\limits_{1}^{n} K_i \subset K$. Then

$$C^*(K_1, ..., K_n) := \bigwedge_{i \in N_n} \bigvee_{\kappa \in K_i} B_{i\kappa} \in \mathcal{B}^*$$

is called a *cube element of* $\mathcal{B}^*$.

The name "cube element" is justified by the following proposition.

7.3.3 Proposition

It holds

$$C^*(K_1, ..., K_n) = B_{\times K_i} \quad (= B_\Gamma \text{ with } \Gamma = \times K_i).$$

Proof

From the distribution law it follows

$$\bigwedge_{i \in N_n} \bigvee_{\kappa \in K_i} B_{i\kappa} = \bigvee_{\substack{j_1 \in K_1 \\ \vdots \\ j_n \in K_n}} \bigwedge_{i \in N_n} B_{ij_i} = \bigvee_{\substack{j_1 \in K_1 \\ \vdots \\ j_n \in K_n}} B_{(j_1, ..., j_n)} = \bigvee_{j \in \times K_i} B_j = B_{\times K_i}.$$

□

7.3.4 Corollary

Suppose $B_j \neq \emptyset$ for each $j \in K$. Then $B_\Gamma \in \mathcal{B}^*$ is a cube element of $\mathcal{B}^*$ if and only if Γ is a cube (see Definition 2.1.1 in case $M = K$).

7.3.5 Definition

A cube element $C^* \in \mathcal{B}^*$ is called an *implicant* of $B_\Gamma \in \mathcal{B}^*$ if $C^* \leq B_\Gamma$. An implicant C^* of B_Γ is called a *prime implicant* of B_Γ if and only if there is no implicant $C^{*\prime}$ of B_Γ with $C^* < C^{*\prime}$ (i.e. $C^* \leq C^{*\prime}$ and $C^* \neq C^{*\prime}$).

7.3.6 Definition

Let $C_1^*, ..., C_r^*$ be implicants (prime implicants) of B_Γ with

$$B_\Gamma = \bigvee_{\rho \in N_r} C_\rho^* \tag{7.3.7}$$

Then (7.3.7) is called a *representation* of B_Γ by *implicants (prime implicants).*

7.3.8 Definition

Let $C_1^*, ..., C_r^*$ be implicants (prime implicants) of B_Γ with (7.3.7) but

$$B_\Gamma \neq \bigvee_{\rho \in N} C_\rho^* \quad \text{for each } N \subset N_r.$$

Then (7.3.7) is called a *reduced* representation of B_Γ by implicants (prime implicants).

7.3.9 Definition

If (7.3.7) holds and if $s \geq r$ for each representation

$$B_\Gamma = \bigvee_{\rho \in N_s} C_\rho^{*\prime}$$

of B_Γ by implicants (prime implicants) then we call (7.3.7) a *minimal* representation of B_Γ by implicants (prime implicants).

7.3.10 Definition

A prime implicant C^* of B_Γ is called *essential* if $C^* \in \{C_1^*, ..., C_r^*\}$ for each representation (7.3.7) of B_Γ by prime implicants.

Now we may formulate the main theorem stating the connection between Boolean elements $B_\Gamma \in \mathcal{B}^*$ and their representations by implicants on the one hand and indicators 1_Γ and their representations by implicants (cube indicators) on the other hand.

7.3.11 Theorem

Suppose $B_j \neq \emptyset$ for all $j \in K = \times \{0, ..., k_i\}$.
Let $M := K$ i.e. $M_i = \{0, ..., k_i\}$ for $i = 1, ..., n$ in Chapter 2. Then for each $\Gamma \subset K$ the following hold.

(a) A cube element $C^*(K_1, ..., K_n)$ of $\mathcal{B}^*$ is an implicant (prime implicant, essential prime implicant) of $B_\Gamma \in \mathcal{B}^*$ if and only if the cube indicator $C(K_1, ..., K_n)$ (see Definition 2.1.1) is an implicant (prime implicant, essential prime implicant) of 1_Γ.

(b) The Boolean element $B_\Gamma \in \mathcal{B}^*$ has the representation (reduced representation, minimal representation) (7.3.7) by the implicants (prime implicants)

$C_1^*, ..., C_r^*$ with $C_\rho^* := C^*(K_{\rho 1}, ..., K_{\rho n})$ for $\rho \in N_r$ if and only if the indicator 1_Γ has the representation (reduced representation, minimal representation)

$$1_\Gamma = \max_{\rho \in N_r} C_\rho$$

by the implicants (prime implicants) $C_1, ..., C_r$ with $C_\rho := C(K_{\rho 1}, ..., K_{\rho n})$ for $\rho \in N_r$.

Proof

(a) Due to Proposition 7.3.3 and Definition 2.1.1 we have

$$C^* := C^*(K_1, ..., K_n) = B_{\times K_i}, \; C := C(K_1, ..., K_n) = 1_{\times K_i}.$$

1. Suppose C^* to be an implicant of B_Γ. Then $B_{\times K_i} \leq B_\Gamma$ and so $\times K_i \leq \Gamma$ according to property 7.2.5 (l). Thus $C = 1_{\times K_i} \leq 1_\Gamma$, i.e. C is an implicant of 1_Γ. Suppose C to be an implicant of 1_Γ, i.e. $1_{\times K_i} \leq 1_\Gamma$. Then $\times K_i \leq \Gamma$ and so $B_{\times K_i} \leq B_\Gamma$ according to property (l) again. Thus C^* is an implicant of B_Γ.

2. Suppose C^* to be a prime implicant of B_Γ Then C is an implicant of 1_Γ due to 1. Assume C not to be a prime implicant of 1_Γ. Then there is an implicant $C' = 1_{\times K_i'}$ of 1_Γ with $C < C'$ and so $\times K_i < \times K_i'$. Then $B_{\times K_i'}$ is an implicant of B_Γ due to 1 with $B_{\times K_i} < B_{\times K_i'}$ due to property (l) and (m). Thus $C^* = B_{\times K_i}$ cannot be a prime implicant of B_Γ in contradiction to the supposition.

Suppose C to be a prime implicant of 1_Γ. Then as above it follows that C^* is a prime implicant of B_Γ.

3. The last proposition (concerning essential prime implicants) will be proved in (c).

(b) 1. Suppose B_Γ to have the indicated representation

$$B_\Gamma = \bigvee_{\rho \in N_r} C_\rho^*$$

by the implicants (prime implicants) $C_1^*, ..., C_r^*$. Since $C_\rho^* = B_{\underset{i}{\times} K_{\rho i}}$ we obtain

$$B_\Gamma = B_{\underset{\rho}{\bigcup} \underset{i}{\times} K_{\rho i}},$$

thus due to property (m)

$$\Gamma = \bigcup_{\rho \in N_r} \mathop{\times}_{i \in N_n} K_{\rho i}$$

and so

$$1_\Gamma = \max_{\rho \in N_r} 1_{\mathop{\times}\limits_i K_{\rho i}} = \max_{\rho \in N_r} C_\rho .$$

Thus 1_Γ has the proposed representation, further $C_1, ..., C_r$ are implicants (prime implicants) of 1_Γ by (a).

2. Suppose 1_Γ to have the representation

$$1_\Gamma = \max_{\rho \in N_r} C_\rho$$

by implicants (prime implicants). Then as above it follows the representation

$$B_\Gamma = \bigvee_{\rho \in N_r} C_\rho^*$$

by implicants (prime implicants).

3. Suppose the representation

$$B_\Gamma = \bigvee_{\rho \in N_r} C_\rho^*$$

to be reduced (minimal). Then it is easy to see that the assumption

$$1_\Gamma = \max_{\rho \in N_r} C_\rho$$

not to be reduced (minimal) yields that

$$B_\Gamma = \bigvee_{\rho \in N_r} C_\rho^*$$

cannot be reduced (minimal) in contradiction to the supposition.

In the same way we obtain that

$$B_\Gamma = \bigvee_{\rho \in N_r} C_\rho^*$$

is reduced (minimal) if

$$1_\Gamma = \max_{\rho \in N_r} C_\rho$$

is it.

(c) It remains to prove that C^* is an essential prime implicant of B_Γ if and only if C is an essential prime implicant of 1_Γ.

1. Suppose C^* to be an essential prime implicant of B_Γ and assume C not to be an essential prime implicant of 1_Γ.

Then 1_Γ has a representation

$$1_\Gamma = \max_{\rho \in N_r} C_\rho$$

by prime implicants with $C \neq C_\rho$ for $\rho \in N_r$. But then B_Γ has a representation

$$B_\Gamma = \bigvee_{\rho \in N_r} C^*_\rho$$

by prime implicants with $C^* \neq C^*_\rho$ for $\rho \in N_r$ in contradiction to the supposition. Thus C is an essential prime implicant of 1_Γ.

2. Suppose C to be an essential prime implicant of 1_Γ. Then as above it follows that C^* is an essential prime implicant of B_Γ. □

Comment

Theorem 7.3.11, (a) says that the problem to gain all implicants (prime implicants, essential prime implicants) of any $B_\Gamma \in \mathcal{B}^*$ is totally solved if we determine all implicants (prime implicants, essential prime implicants) of the indicator 1_Γ. And this we can do by the methods of Chapter 2.

The second part of Theorem 7.3.11 moreover states that we obtain all kinds of representations of B_Γ by implicants (prime implicants) in a simple way from the corresponding representations of 1_Γ by implicants (prime implicants) treated also in Chapter 2.

7.4 Representation of Boolean Elements by Implicates

Now let us consider Boolean implicates of Boolean elements out of $\mathcal{B}^*$. We may regard them as generalizations of the indicator implicants of Chapter 3. We use the notation of Section 7.3.

7.4.1 Definition

Let $\phi \neq \times K_i \subset K$. Then

$$F^*(K_1, ..., K_n) := \bigvee_{i \in N_n} \bigvee_{\kappa \in K_i} B_{i\kappa}$$

is called an *anticube element of* $\mathcal{B}^*$. The name "anticube element" is justified by the following proposition.

7.4.2 Proposition

It holds

$$F^*(K_1, ..., K_n) = \neg C^*(\overline{K}_1, ..., \overline{K}_n) = B_{K \setminus \times \overline{K}_i} \quad (\text{with } \overline{K}_i := \{0, ..., k_i\} \setminus K_i).$$

Proof

From de Morgan's rule it follows $\neg F^*(K_1, ..., K_n) = \bigwedge_{i \in N_n} (\neg \bigvee_{k \in K_i} B_{ik})$.
Since $\{B_{i0}, ..., B_{ik_i}\}$ is a partition of $\mathbb{1}$, we obtain $\neg \bigvee_{k \in K_i} B_{ik} = \bigvee_{k \in \overline{K}_i} B_{ik}$
and so $\neg F^*(K_1, ..., K_n) = \bigwedge_{i \in N_n} \bigvee_{k \in \overline{K}_i} B_{ik} = C^*(\overline{K}_1, ..., \overline{K}_n) = B_{\times \overline{K}_i}$ (according to Proposition 7.2.3). From this the proposition follows by Theorem 7.2.3, (e).

Remark

For any cube $\times_1^n K_i \subseteq K$ define for $i = 0, ..., n$ the cubes

$$K_i^* := \{j \in K : j_i \in K_i\} = \times_1^n L_r \quad \text{with } L_r = \begin{cases} K_i & \text{for } r = i \\ \{0, ..., k_r\} & \text{for } r \neq i \,. \end{cases}$$

Then

$$K \setminus \times_1^n \overline{K}_i = K \setminus \bigcap_1^n \overline{K_i^*} = \bigcup_1^n K_i^*.$$

Thus we may write

$$F^*(K_1, ..., K_n) = B_{\bigcup K_i^*} = \bigvee_{i \in N_i} B_{K_i^*}. \tag{7.4.3}$$

7.4.4 Corollary

Suppose $B_j \neq \mathbb{O}$ for each $j \in K$. Then $B_\Gamma \in \mathcal{B}^*$ is an anticube element of $\mathcal{B}^*$ if and only if Γ is an anticube, i.e. $\Gamma = \bigcup K_i^*$ (see Definition 3.1.1 in case $M = K$).

Proof

This follows from (7.4.3) and Lemma 7.2.4, d). □

The following definitions concern implicates and representations of Boolean elements out of $\mathcal{B}^*$ by implicates. They are the complete counterparts to the Definitions 7.3.5, ..., 7.3.10.

7.4.5 Definition

An anticube element $F^* \in \mathcal{B}^*$ is called an *implicate* of $B_\Gamma \in \mathcal{B}^*$ if $B_\Gamma \leq F^*$. An implicate F^* of B_Γ is called a *prime implicate* of B_Γ if there is no implicate $F^{*'}$ of B_Γ with $F^{*'} < F^*$ (i.e. $F^{*'} \leq F^*$ and $F^{*'} \neq F^*$).

7.4.6 Definition

Let $F_1^*, ..., .F_r^*$ be implicates (prime implicates) of B_Γ with

$$B_\Gamma = \bigwedge_{\rho \in N_\rho} F_\rho^*. \tag{7.4.7}$$

Then (7.4.7) is called a *representation* of B_Γ by *implicates (prime implicates).*

7.4.8 Definition

Let $F_1^*, ..., F_r^*$ be implicates (prime implicates) of B_Γ with (7.4.7) but

$$B_\Gamma \neq \bigwedge_{\rho \in N} F_\rho^* \quad \text{for each} \quad N \subset N_r.$$

Then (7.4.7) is called a *reduced* representation of B_Γ by implicates (prime implicates).

7.4.9 Definition

If (7.4.7) holds and if $s \geq r$ for each representation

$$B_\Gamma = \bigwedge_{\rho \in N_s} F_\rho^{*'}$$

of B_Γ by implicates (prime implicates) then we call 7.4.7 a *minimal* representation of B_Γ by implicates (prime implicates).

7.4.10 Definition

A prime implicate F^* of B_Γ is called *essential* if $F^* \in \{F_1^*, ..., .F_r^*\}$ for each representation (7.4.7) of B_Γ by prime implicates.

The following main theorem now states the strong connection between Boolean elements $B_\Gamma \in \mathcal{B}^*$ and their representations by implicates on the one hand and indicators 1_Γ and their representation by implicates (anticube indicators) on the other hand. The analogy to Theorem 7.3.11 is obvious.

7.4.11 Theorem

Suppose $B_j \neq \mathcal{O}$ for all $j \in K$ and let $M := K$. For any $\Gamma \subset K$ then the following hold.

(a) An anticube element $F^*(K_1, ..., K_n)$ of $\mathcal{B}^*$ is an implicate (prime implicate, essential prime implicate) of $B_\Gamma \in \mathcal{B}^*$ if and only if the anticube indicator $F(K_1, ..., K_n)$ (see Definition 3.1.1) is an implicate (prime implicate, essential prime implicate) of 1_Γ.

(b) The Boolean element $B_\Gamma \in \mathcal{B}^*$ has the representation (reduced representation, minimal representation)

$$B_\Gamma = \bigwedge_{\rho \in N_r} F^*_\rho$$

by the implicates (prime implicates) $F^*_1, ..., F^*_r$ with $F^*_\rho := F^*(K_{\rho 1}, ...K_{\rho n})$ for $\rho \in N_r$ if and only if the indicator 1_Γ has the representation (reduced representation, minimal representation)

$$1_\Gamma = \min_{\rho \in N_r} F_\rho = \prod_{\rho \in N_r} F_\rho$$

by the implicates (prime implicates) $F_1, ..., F_r$ with $F_\rho := F(K_{\rho 1}, ..., K_{\rho n})$ for $\rho \in N_r$.

Proof

We omit the proof since it is similar to 7.3.11.

7.5 Probability

From a Boolean algebra $\mathcal{B}$ we obtain a probability space by introduction of a probability measure on the elements of $\mathcal{B}$. In particular we are interested in probability measures on Boolean algebras of type $\mathcal{B}^*$.

7.5.1 Definition

Let $\mathcal{B}$ be a Boolean algebra with 0–element $\mathcal{O}$ and 1–element $\mathbb{1}$. A function $P : \mathcal{B} \to [0,1]$ is called a *probability measure* – shortly a *probability* – if the following holds:

(a) $P(\mathbb{1}) = 1$,
(b) $B, B' \in \mathcal{B}, B \wedge B' = \mathbb{O} \Rightarrow P(B \vee B') = P(B) + P(B')$.

Remark

From (a) and (b) we obtain

(c) $P(\mathbb{O}) = 0$.

From (b) we obtain (by induction)

(b') $B^{(1)}, ..., B^{(r)} \in \mathcal{B}, B^{(i)} \wedge B^{(j)} = \mathbb{O}$ for $i \neq j$
$\Rightarrow P(B^{(1)} \vee ... \vee B^{(r)}) = P(B^{(1)}) + ... + P(B^{(r)})$.

We are not interested in a so-called σ-additivity.

A probability on $\mathcal{B}^*$ may be characterized by the following statement.

7.5.2 Theorem

Let $\{B_{10}, ..., B_{1k_1}\}, ..., \{B_{n0}, ..., B_{nk_n}\}$ be n partitions of the 1-element $\mathbb{1}$ of a Boolean Algebra $\mathcal{B}$ and $\mathcal{B}^* := \mathcal{B}(\{B_{10}, ..., B_{nk_n}\}) = \{B_\Gamma : \Gamma \subseteq K\}$ the Boolean algebra generated by $\{B_{10}, ..., B_{nk_n}\}$ according to Theorem 7.2.3.

For $j \in K$ let

$$p_j \begin{cases} \geq 0 & \text{if } B_j \neq \mathbb{O} \\ = 0 & \text{if } B_j = \mathbb{O}. \end{cases}$$

Then for $P : \mathcal{B}^* \to \mathbb{R}$ the following are equivalent:

(a) P is a probability on $\mathcal{B}^*$ with $P(B_j) = p_j$ for $j \in K$.
(b) $\sum_{j \in K} p_j = 1$, $P(B_\Gamma) = \sum_{j \in \Gamma} p_j$ for $\Gamma \subseteq K$.

Proof

Suppose (a). Since $B_j \wedge B_k = \mathbb{O}$ for $j \neq k$ we obtain

$$\sum_{j \in \Gamma} p_j = \sum_{j \in \Gamma} P(B_j) = P(\bigvee_{j \in \Gamma} B_j) = P(B_\Gamma),$$

in particular

$$\sum_{j \in K} p_j = P(B_K) = P(\mathbb{1}) = 1.$$

Suppose (b). Then

$$P(\mathbb{1}) = P(B_K) = \sum_{j \in K} p_j = 1.$$

Now let $B_\Gamma \wedge B_{\Gamma'} = \mathbb{O}$. Due to Theorem 7.2.3, d) it holds $B_{\Gamma \cap \Gamma'} = \mathbb{O}$ implying $B_j = \mathbb{O}$ for $j \in \Gamma \cap \Gamma'$. From Theorem 7.2.3, c) we obtain $P(B_\Gamma \vee B_{\Gamma'}) =$

$P(B_{\Gamma\cup\Gamma'}) = \sum_{j\in\Gamma\cup\Gamma'} p_j = \sum_{j\in\Gamma} p_j + \sum_{j\in\Gamma'} p_j - \sum_{j\in\Gamma\cap\Gamma'} p_j = P(B_\Gamma) + P(B_{\Gamma'})$, since $p_j = 0$ for $j \in \Gamma\cap\Gamma'$.

Furthermore

$$P(B_j) := P(B_{\{j\}}) = \sum_{i\in\{j\}} p_i = p_j.$$

Finally $0 \leq \sum_{j\in\Gamma} p_j \leq 1$ and so $P(B_\Gamma) \in [0,1]$. □

It is desirable to state a probability on $\mathcal{B}^*$ such that the generating elements have given probability values. In fact we obtain such a probability in a unique way by introduction of the stochastic independence of the generating partitions of $\mathbb{1}$.

7.5.3 Definition

Let P be a probability on $\mathcal{B}^*$.

The n partitions $\{B_{10}, ..., B_{1k_1}\}, ..., \{B_{n0}, ..., B_{nk_n}\}$ of $\mathbb{1}$ are called (stochastically) *independent* (with respect to the probability P) if and only if for all $B_{1j_1}, ..., B_{nj_n}$ with $j_1 \in \{0, ..., k_1\}, ..., j_n \in \{0, ..., k_n\}$ (i.e. $j := (j_1, ..., j_n) \in K$) holds

$$P(\bigwedge_{i\in N_n} B_{ij_i}) = \prod_{i\in N_n} P(B_{ij_i}).$$

Remark

For any partition $\{B^{(1)}, ..., B^{(r)}\}$ of $\mathbb{1}$ the Boolean algebra $\mathcal{B}(\{B^{(1)}, ..., B^{(r)}\})$, generated by $\{B^{(1)}, ..., B^{(r)}\}$ is given by

$$\mathcal{B}(\{B^{(1)}, ..., B^{(r)}\}) = \{\bigvee_{\rho\in N} B^{(\rho)} : N \subseteq N_r\}$$

with $\mathbb{O} := \bigvee_{\rho\in\emptyset} B^{(\rho)}$, $\mathbb{1} = \bigvee_{\rho\in N_r} B^{(\rho)}$.

For a probability P on $\mathcal{B}(\{B^{(1)}, ..., B^{(r)}\})$ holds

$$P(\bigvee_{\rho\in N} B^{(\rho)}) = \sum_{\rho\in N} P(B^{(\rho)}).$$

Next we state an equivalent characterization of independence.

7.5.4 Theorem

Define for $i \in N_n$

$$\mathcal{B}_i^* := \mathcal{B}(\{B_{i0}, ..., B_{ik_i}\})$$

(Boolean algebra generated by the partition $\{B_{i0}, ..., B_{ik_i}\}$).

Then the following are equivalent:

(a) The n partitions $\{B_{10}, ..., B_{1k_1}\}, ..., \{B_{n0}, ..., B_{nk_n}\}$ are independent.

(b) For all $D_1, ..., D_n \in \mathcal{B}^*$ with $D_i \in \mathcal{B}_i^*$ for $i \in N_n$ holds

$$P(\bigwedge_{i \in N_n} D_i) = \prod_{i \in N_n} P(D_i).$$

Proof

This follows from the distribution law by using the partition properties (see also the Remark to Definition 7.5.3).

Now we state the existence of a probability on $\mathcal{B}^*$ with given probability values for the elements $B_{10}, ..., B_{nk_n}$.

7.5.5 Theorem

Let $p_{10}, ..., p_{1k_1}, ..., p_{n0}, ..., p_{nk_n} \geq 0$ and $\sum_{k=0}^{k_i} p_{ik} = 1$ for $i \in N_n$. Then the following holds:

1. Suppose $B_j \neq \emptyset$ for $j \in K$ (see Theorem 7.2.3). Then there is exactly one probability P on $\mathcal{B}^*$ with

   (*) $$P(B_j) = \prod_{i \in N_n} p_{ij_i} \text{ for } j \in K.$$

2. Let P be a probability on $\mathcal{B}^*$. Then the following are equivalent:

   (a) (*) holds.

   (b) $P(B_{10}) = p_{10}, ..., P(B_{nk_n}) = p_{nk_n}$ and $P(B_\Gamma) = \sum_{j \in \Gamma} P(B_j) = \sum_{j \in \Gamma} \prod_{i \in N_n} p_{ij_i}$ for $\Gamma \subseteq K$.
   The generating partitions $\{B_{10}, ..., B_{1k_1}\}, ..., \{B_{n0}, ..., B_{nk_n}\}$ are independent.

Proof

1. Suppose $B_j \neq \emptyset$ for $j \in K$. If there exists a probability P with (*) then it holds

(**) $$P(B_\Gamma) = \sum_{j \in \Gamma} \prod_{i \in N_n} p_{ij_i} \quad \text{for } \Gamma \subseteq K.$$

We still have to show that P satisfying (**) is a probability.

α) By definition it holds $P(B_\Gamma) \geq 0$.

β) $P(\mathbb{1}) = P(B_K) = \sum\limits_{j \in K} \prod\limits_{i \in N_n} p_{ij_i} = \prod\limits_{i \in N_n} \sum\limits_{j_i=0}^{k_i} p_{ij_i} = \prod\limits_{i \in N_n} 1 = 1.$

γ) Let $B_\Gamma \wedge B_{\Gamma'} = \mathbb{O}$. Then $\Gamma \cap \Gamma' = \emptyset$ and
$P(B_\Gamma \vee B_{\Gamma'}) = P(B_{\Gamma \cup \Gamma'}) = \sum\limits_{j \in \Gamma \cup \Gamma'} \prod\limits_{i \in N_n} p_{ij_i}$
$= \sum\limits_{j \in \Gamma} \prod\limits_{i \in N_n} p_{ij_i} + \sum\limits_{j \in \Gamma'} \prod\limits_{i \in N_n} p_{ij_i} = P(B_\Gamma) + P(B_{\Gamma'})$.
Thus P is a probability.

2. Suppose (a). The representation of $P(B_\Gamma)$ is obvious (see 1).
For $i_0 \in N_n, j_0 \in \{0, ..., k_{i0}\}$ it holds (see the proof of Theorem 7.2.3)

$$B_{i_0 j_0} = B_{\Gamma'} = \bigvee_{j \in \Gamma'} B_j \quad \text{with } \Gamma' := \{(j_1, ..., j_n) \in K : j_{i_0} = j_0\}.$$

This implies

$$P(B_{i_0 j_0}) = \sum_{j \in \Gamma'} P(B_j) = \sum_{j \in \Gamma'} \prod_{i \in N_n} p_{ij_i} = \sum_{\substack{j_i \in \{0,...,k_i\} \\ \text{for } i \in N_n \setminus \{i_0\}}} \left[p_{i_0 j_0} \prod_{i \in N_n \setminus \{i_0\}} p_{ij_i} \right]$$

$$= p_{i_0 j_0} \prod_{i \in N_n \setminus \{i_0\}} \sum_{j_i=0}^{k_i} p_{ij_i} = p_{i_0 j_0} \prod_{i \in N_n \setminus \{i_0\}} 1 = p_{i_0 j_0}.$$

Thus $P(B_{10}) = p_{10}, ..., P(B_{nk_n}) = p_{nk_n}$.
Finally for each $j = (j_1, ..., j_n) \in K$ we obtain

$$P(\bigwedge_{i \in N_n} B_{ij_i}) = P(B_j) = \prod_{i \in N_n} p_{ij_i},$$

i.e. the n partitions are independent.
Now suppose (b). Then (*) follows by the definition of independence. □

In the following we suppose the independence of the generating partitions. In this case we obtain simplified expressions for the probability of cube elements and anticube elements.

7.5.6 Theorem

Let P be a probability on $\mathcal{B}^*$ with independent generating partitions $\{B_{10}, ..., B_{1k_1}\}, ..., \{B_{n0}, ..., B_{nk_n}\}$ satisfying $P(B_{ij_i}) = p_{ij_i}$ for $i \in N_n, j_i \in \{0, ..., k_i\}$.

Let $C^*(K_1, ..., K_n)$ be a cube element of $\mathcal{B}^*$ (see Definition 7.3.2). Then

(7.5.7)
$$P(C^*(K_1, ..., K_n)) = \prod_{i \in N_n} (\sum_{\kappa \in K_i} p_{i\kappa}).$$

Proof

From Definition 7.3.2 we have

$$C^*(K_1, ..., K_n) = \bigwedge_{i \in N_n} D_i$$

with

$$D_i = \bigvee_{\kappa \in K_i} B_{i\kappa} \in \mathcal{B}_i^* := \mathcal{B}(\{B_{i0}, ..., B_{ik_i}\})$$

and

$$P(D_i) = \sum_{\kappa \in K_i} P(B_{i\kappa}) = \sum_{\kappa \in K_i} p_{i\kappa}$$

for $i \in N_n$ (see also the remark to Definition 7.5.3). Now Theorem 7.5.4 yields

□
$$P(C^*(K_1, ..., K_n)) = \prod_{i \in N_n} P(D_i) = \prod_{i \in N_n} (\sum_{\kappa \in K_i} p_{i\kappa}).$$

7.5.8 Corollary

Let P be given as in Theorem 7.5.6 and $F^*(K_1, ..., K_n)$ be an anticube element of $\mathcal{B}^*$ (see Definition 7.4.1). Then

$$\begin{aligned} P(F^*(K_1, ..., K_n)) &= 1 - \prod_{i \in N_n} (\sum_{\kappa \in \overline{K}_i} p_{i\kappa}) \\ &= 1 - \prod_{i \in N_n} (1 - \sum_{\kappa \in K_i} p_{i\kappa}). \end{aligned}$$

Proof

The assertion follows from Proposition 7.4.2 and Theorem 7.5.6.

If an element B_Γ of $\mathcal{B}^*$ has a representation by "mutually disjoint" implicants then the probability of B_Γ is given by the sum of their probabilities, i.e. we may state the following result.

7.5.9 Theorem

Let $C_1^*, \dots, C_r^*$ be implicants of B_Γ with $C_j^* \wedge C_{j'}^* = \emptyset$ for $j \neq j'$ and

$$B_\Gamma = \bigvee_{\rho \in N_r} C_\rho^*.$$

Then

$$P(B_\Gamma) = \sum_{\rho \in N_r} P(C_\rho^*),$$

where $P(C_\rho^*)$ for $\rho \in N_r$ is given according to Theorem 7.5.6.

Proof

The assertion is evident.

The following is the implicate counterpart of Theorem 7.5.9.

7.5.10 Theorem

Let $F_1^*, \dots, F_r^*$ be implicates of B_Γ with $F_j^* \vee F_{j'}^* = \mathbb{1}$ for $j \neq j'$ and

$$B_\Gamma = \bigwedge_{\rho \in N_r} F_\rho^*.$$

Then

$$P(B_\Gamma) = 1 - \sum_{\rho \in N_r} (1 - P(F_\rho^*)),$$

where $P(F_\rho^*)$ for $\rho \in N_r$ is given according to Corollary 7.5.8.

Proof

The assertion follows from

$$\neg B_\Gamma = \bigvee_{\rho \in N_r} (\neg F_\rho^*),$$

since $F_j^* \vee F_{j'}^* = \mathbb{1}$ is equivalent to $(\neg F_j^*) \wedge (\neg F_{j'}^*) = \emptyset$.

Comment

We showed in this section that any Boolean algebra of type $\mathcal{B}^*$ (i.e. generated by n partitions of the 1–element $\mathbb{1}$) may be provided with a probability P. Especially we considered the case where the n partitions are independent with given probabilities for the generating elements $B_{10}, \dots, B_{nk_n}$.

In Chapter 8 several Boolean algebras of type $\mathcal{B}^*$ will be presented: set algebras, indicator algebras, algebras of classes of propositions and truth function algebras.

Obviously we may use the results of Section 7.5 to state a probability P and to calculate the probabilities of the elements respectively.

We leave it to the reader to translate the results of Section 7.5 into the corresponding versions concerning sets, indicators etc.

Chapter 8

Applications

8.1 Set Algebras (Event Algebras)

Set algebras in a set Ω often are interpreted in probability theory as systems of events, where the empty set ϕ means the impossible event, Ω means the sure event. The event $A \cup B$ occurs if at least one of the events A and B occurs, the event $A \cap B$ occurs if both the event A and the event B occur. The event $\overline{A}$ occurs if A does not occur (conf. e.g. Wilks (1962)).

In the following we consider set algebras generated by n partitions of the set Ω, as described in the beginning of Chapter 7. The explicite representation of such algebras is given by the following statement.

8.1.1 Theorem

Let Ω be a set and $\{A_{10}, ..., A_{1k_1}\}, ..., \{A_{n0}, ..., A_{nk_n}\}$ be n partitions of Ω. Then the set algebra $\mathcal{A}(\{A_{10}, ..., A_{nk_n}\})$ in Ω generated by $\{A_{10}, ..., A_{nk_n}\}$ is given by

$$\mathcal{A}^* := \mathcal{A}(\{A_{10}, ..., A_{nk_n}\}) = \{A_\Gamma : \Gamma \subseteq K\}$$

with

$$K := \bigtimes_{i \in N_n} \{0, ..., k_i\},$$
$$A_j := \bigcap_{i \in N_n} A_{ij_i} \text{ for each } j := (j_1, ..., j_n) \in K$$
$$A_\Gamma := \bigcup_{j \in \Gamma} A_j \text{ for each } \Gamma \subseteq K, \ A_\emptyset := \emptyset.$$

Further it holds

$$A_K = \Omega, A_\Gamma \cup A_{\Gamma'} = A_{\Gamma \cup \Gamma'}, A_\Gamma \cap A_{\Gamma'} = A_{\Gamma \cap \Gamma'}, \overline{A_\Gamma} := \Omega \setminus A_\Gamma = A_{K \setminus \Gamma}.$$

The nonempty sets A_j are the atoms of $\mathcal{A}^*$. The atoms form a partition of Ω.

The set algebra $\mathcal{A}^*$ is the Boolean algebra generated by $\{A_{10}, ..., A_{nk_n}\}$ with $A \leq A' \Leftrightarrow A \subseteq A', \mathcal{O} = \emptyset, \mathbb{1} = \Omega, A \vee A' = A \cup A', A \wedge A' = A \cap A', \neg A = \overline{A} = \Omega \setminus A$.

Proof

Theorem 7.1.5 with $\mathcal{B} = \mathcal{A} = \mathcal{P}(\Omega)$ yields that $\mathcal{P}(\Omega)$ is a Boolean algebra with the order relation $\subseteq$ and $\mathcal{O} = \emptyset, \mathbb{1} = \Omega, A \vee A' = A \cup A', A \wedge A' = A \cap A', \neg A = \overline{A} := \Omega \setminus A$ (see also the remark to Definition 7.1.3).

Now the proposition follows from Theorem 7.2.3 . □

8.1.2 Properties of $\mathcal{A}^* = \mathcal{A}(\{A_{10}, ..., A_{nk_n}\}) = \{A_\Gamma : \Gamma \subseteq K\}$

Suppose $A_j \neq \emptyset$ for all $j \in K$. Then we may read the list 7.2.5 of properties of $\mathcal{B}(\{B_{10}, ..., B_{nk_n}\})$ as a list of properties of $\mathcal{A}^*$. We only have always to replace B by A, $\mathbb{O}$ by $\emptyset$, $\mathbb{1}$ by Ω, $\vee$ by $\cup$, $\wedge$ by $\cap$, $\neg B$ by $\overline{A}$ and $\leq$ by $\subseteq$.

8.1.3 Representation of Sets out of $\mathcal{A}^*$ by Implicants

Since $\mathcal{A}^*$ is the Boolean algebra generated by $\{A_{10}, ..., A_{nk_n}\}$, we may translate the ideas of implicants and representations of implicants of $\mathcal{B}^*$ to $\mathcal{A}^*$.

8.1.3.1 Definition

Let $\emptyset \neq \times K_i \subset K$. Then

(8.1.3.2)
$$C^*(K_1, ..., K_n) := \bigcap_{i \in N_n} \bigcup_{k \in K_i} A_{ik} = A_{\times K_i} \in \mathcal{A}^*$$

is called a *cube element of* $\mathcal{A}^*$.

Remark (see Proposition 7.3.3 and Corollary 7.3.4)

Let $M = K$ and suppose $A_j \neq \emptyset$ for each $j \in K$. Then any $A_\Gamma \in \mathcal{A}^*$ is a cube element if and only if Γ is a cube according to Definition 2.1.1.

We may interprete $C^*(K_1, ..., K_n)$ as an event by

"The event $C^*(K_1, ..., K_n)$ occurs if and only if for each $i \in N_n$ exactly one of the events A_{ik} with $k \in K_i$ occur."

8.1.3.3 Definition

A cube element $C^* \in \mathcal{A}^*$ is called an *implicant* of $A_\Gamma \in \mathcal{A}^*$ if $C^* \subseteq A_\Gamma$. An implicant C^* of A_Γ is called a *prime implicant* of A_Γ if there is no implicant $C^{*'}$ of A_Γ with $C^* \subset C^{*'}$.

8.1.3.4 Definition

Let $C_1^*, ..., C_r^*$ be implicants (prime implicants) of A_Γ with

(8.1.3.5)
$$A_\Gamma = \bigcup_{\rho \in N_r} C_\rho^* .$$

Then (8.1.3.5) is called *a representation of* A_Γ *by implicants (prime implicants).*

Remark

The interpretation of (8.1.3.5) as an event is obvious: "A_Γ occurs if and only if at least one of the "cube events" $C_1^*, ..., C_r^*$ occurs" (see Remark to Definition 8.1.3.1).

The following definitions are the set versions of the corresponding general definitions of Chapter 7.

8.1.3.6 Definition

Reduced representations of A_Γ by implicants (prime implicants) are defined by Definition 7.3.8: Replace B_Γ by A_Γ,(7.3.7) by (8.1.3.5).

8.1.3.7 Definition

Minimal representations of A_Γ by implicants (prime implicants) are defined by Definition 7.3.9: Replace (7.3.7) by (8.1.3.5), B_Γ by A_Γ.

8.1.3.8 Definition

Essential prime implicants of A_Γ are defined by Definition 7.3.10: Replace B_Γ by A_Γ (7.3.7) by (8.1.3.5).

Next we state the set version of Theorem 7.3.11.

8.1.3.9 Theorem

Suppose $A_j \neq \emptyset$ for all $j \in K$. Let $M = K$. Then for each $\Gamma \subset K$ the following hold.

(a) A cube element $C^*(K_1, ..., K_n)$ of $\mathcal{A}^*$ is an implicant (prime implicant, essential prime implicant) of $A_\Gamma \in \mathcal{A}^*$ if and only if the cube indicator $C(K_1, ..., K_n)$ is an implicant (prime implicant, essential prime implicant) of 1_Γ.

(b) The set $A_\Gamma \in \mathcal{A}^*$ has the representation (reduced representation, minimal representation) (8.1.3.5) by the implicants (prime implicants) $C_1^*, ..., C_r^*$ with $C_\rho^* := C^*(K_{\rho 1}, ..., K_{\rho n})$ (conf. (8.1.3.2)) for $\rho \in N_r$ if and only if the indicator 1_Γ has the representation (reduced representation, minimal representation)

$$1_\Gamma = \max_{\rho \in N_r} C_\rho$$

by the implicants (prime implicants) $C_1, ..., C_r$ with $C_\rho := C(K_{\rho 1}, ..., K_{\rho n})$ for $\rho \in N_r$.

Comment

In Chapter 2 we saw that each indicator 1_Γ has several representations by implicants, e.g. minimal representations by prime implicants.

By Theorem 8.1.3.9 this implies that also each event A_Γ of $\mathcal{A}^*$ has such representations given by (8.1.3.5) with (see (8.1.3.2))

$$C_\rho^* = \bigcap_{i \in N_n} \bigcup_{k \in K_{\rho i}} A_{ik} \,.$$

We may express this by the statement "The event A_Γ occurs if and only if for at least one $\rho \in N_r$ exactly one of the events A_{ik} with $k \in K_{\rho i}$ occurs for all $i \in N_n$".

8.1.4 Representations of Sets out of $\mathcal{A}^*$ by Implicates

Now let us consider representations of a set A_Γ by implicates. First we give the appropriated definitions.

8.1.4.1 Definition

Let $\phi \neq \times K_i \subset K$. Then

(8.1.4.2) $$F^*(K_1, ..., K_n) := \bigcup_{i \in N_n} \bigcup_{k \in K_i} A_{ik} \ \ (= \bigcup_{i \in N_n} \bigcap_{k \in \overline{K_i}} \overline{A_{ik}}) \in \mathcal{A}^*$$

is called an *anticube element of* $\mathcal{A}^*$.

Remark (see Proposition 7.4.2 and Corollary 7.4.4)

It holds

$$F^*(K_1, ..., K_n) = \overline{C^*(\overline{K_1}, ..., \overline{K_n})} = A_{K \setminus \times \overline{K_i}} \,.$$

Set $M = K$ and suppose $A_j \neq \emptyset$ for each $j \in K$. Then $A_\Gamma \in \mathcal{A}^*$ is an anticube element if and only if Γ is an anticube according to Definition 3.1.1.

We may interprete $F^*(K_1, ..., K_n)$ as an event by

"The event $F^*(K_1, ..., K_n)$ occurs if and only if for at least one $i \in N_n$ exactly one of the events A_{ik} with $k \in K_i$ occurs."

And we may $F^*(K_1, ..., K_n)$ also express by

"The event $F^*(K_1, ..., K_n)$ occurs if and only if for at least one $i \in N$ none of the events A_{ik} with $k \in \overline{K_i}$ occurs."

8.1.4.3 Definition

An anticube element $F^* \in \mathcal{A}^*$ is called an *implicate* of $A_\Gamma \in \mathcal{A}^*$ if $A_\Gamma \subseteq F^*$. An implicate F^* of A_Γ is called a *prime implicate* of A_Γ if there is no implicate $F^{*'}$ of A_Γ with $F^{*'} \subset F^*$.

8.1.4.4 Definition

Let $F_1^*, ..., F_r^*$ be implicates (prime implicates) of A_Γ with

$$A_\Gamma = \bigcap_{\rho \in N_r} F_\rho^* . \qquad (8.1.4.5)$$

Then (8.1.4.5) is called a *representation* of A_Γ by implicates (prime implicates).

Remark

The obvious interpretation of (8.1.4.5) is "A_Γ occurs if and only if all "anticube events" $F_1^*, ..., F_r^*$ occur" (see Remark to Definition 8.1.4.1).

8.1.4.6 Definition

As to the definition of reduced (minimal) representations by (prime) implicates as well as the definition of essential prime implicates we refer to the corresponding Definitions 8.1.3.6, 8.1.3.7, 8.1.3.8 (respectively 7.4.8, 7.4.9, 7.4.10).

Now we may state the set version of Theorem 7.4.11, i.e. the implicate analogue of Theorem 8.1.3.9. It concerns the representations of any A_Γ by implicates (prime implicates).

8.1.4.7 Theorem

Suppose $A_j \neq \emptyset$ for all $j \in K$. Let $M = K$. Then for each $\Gamma \subset K$ the following hold.

(a) An anticube element $F^*(K_1, ..., K_n)$ of $\mathcal{A}^*$ is an implicates (prime implicate, essential implicate) of A_Γ if and only if the anticube indicator $F(K_1, ..., K_n)$ (see Definition 3.1.1) is an implicates (prime implicate) of 1_Γ (see Definition 3.2.3).

(b) The set $A_\Gamma \in \mathcal{A}^*$ has the representation (reduced representation, minimal representation) (8.1.4.5) by the implicates (prime implicates $F_1^*, ..., F_r^*$ with $F_\rho^* := F^*(K_{\rho 1}, ..., K_{\rho n})$ (conf. also (8.1.4.2)) for $\rho \in N_r$ if and only if the indicator 1_Γ has the representation (reduced representation, minimal representation)

$$1_\Gamma = \min_{\rho \in N_r} F_\rho = \prod_1^r F_\rho$$

by the implicates (prime implicates) $F_1, ..., F_r$ with $F_\rho = F(K_{\rho 1}, ..., K_{\rho n})$ for $\rho \in N_r$.

Comment

Theorem 2.1.4.7 says that we obtain all representations of any set A_Γ by implicates (prime implicates) immediately from the corresponding representations of 1_Γ, and we may express the representation (8.1.4.5) (with $F_\rho^* = F^*(K_{\rho 1}, ..., K_{\rho n})$ for $\rho \in N_r$) by

"The event A_Γ occurs if and only if for each $\rho \in N_r$ exactly one of the events A_{ik} with $k \in K_{\rho i}$ occurs for at least one $i \in N_n$".

8.1.5 Algebras of Type $\mathcal{A}^*$ with K Atoms

It seems obvious to ask for the existence of a set Ω with n partitions $\{A_{10}, ..., A_{1k_1}\}, ..., \{A_{n0}, ..., A_{nk_n}\}$ with $A_j := \bigcap_{i \in N_n} A_{ij_i} \neq \emptyset$ for all $j \in K$. In this case from Theorem 8.1.1 we know that the sets $A_j, j \in K$ are the atoms of $\mathcal{A}^*$. Clearly each set Ω' has at least one partition into any r nonempty subsets if and only if $|\Omega'| \geq r$.

Now for $i \in N_n$ let Ω_i be a set with $|\Omega_i| \geq k_i$ and $\{A'_{i0}, ..., A'_{ik_i}\}$ a partition of Ω_i. Define

$$\Omega := \bigtimes_1^n \Omega_i$$

and for $i \in N_n, \kappa \in \{0, ..., k_i\}$,

$$A_{i\kappa} := \{(x_1, ..., x_n) \in \Omega : x_i \in A'_{i\kappa}\}, \kappa = 0, ..., k_i.$$

Then for $i \in N_n$,

$$A_{i\kappa} \neq \emptyset \quad \text{for } \kappa = 0, ..., k_i \quad \text{and} \quad \bigcup_{\kappa=0}^{k_i} A_{i\kappa} = \Omega.$$

Thus $\{A_{10}, ..., a_{1k_1}\}, ..., \{A_{n0}, ..., A_{nk_n}\}$ are n partitions of Ω. Further for any $j := (j_1, ..., j_n) \in K$ holds

$$A_j := \bigcap_{i \in N_n} A_{ij_i} = \{(x_1, ..., x_n) \in \Omega : x_i \in A'_{ij_i} \text{ for } i = 1, ..., n\}$$

$$= \bigtimes_{i=1}^n A'_{ij_i} \neq \emptyset.$$

Examples

1. Subsets of M (see Chapter 2)

$$\Omega_i := M_i := \{a_{i0}, ..., a_{ik_i}\}, i \in N_n,$$
$$A'_{i\kappa} := \{a_{i\kappa}\}, \kappa \in \{0, ..., k_i\},$$
$$\Omega := \times M_i =: M,$$
$$A_{i\kappa} := \{(x_1, ..., x_n) \in M : x_i = a_{i_\kappa}\}, i \in N_n, \kappa \in \{0, ..., k_i\},$$
$$A_j := \bigcap_{i \in N_n} A_{ij_i} = \times_{i=1}^{n} A'_{ij_i} = \times_{i=1}^{n} \{a_{ij_i}\} = \{a_j\} := \{(a_{1j_1}, ..., a_{nj_n})\} \neq \emptyset$$
for $j := (j_1, ..., j_n) \in K$.

This means: The set M, defined in Chapter 2, has the n partitions $\{A_{10}, ..., A_{1k_1}\}, ..., \{A_{n0}, ..., A_{nk_n}\}$ given as above. Theorem 8.1.1 now says that the set algebra in M, generated by $\{A_{10}, ..., A_{nk_n}\}$ is given by the system

$$\mathcal{A}^* = \{A_\Gamma : \Gamma \subseteq K\}$$

with

$$A_\Gamma := \bigcup_{j \in \Gamma} A_j = \bigcup_{j \in \Gamma} \{a_j\} = \{a_j : j \in \Gamma\} =: G_\Gamma \quad \text{for } \Gamma \subseteq K,$$

so that

$$\mathcal{A}^* = \mathcal{P}(M),$$

(see the Preliminary Remark in Chapter 2).

In fact $\mathcal{A}^*$ is the system of all subsets of M, and the one–point–sets $\{a\}$ with $a \in M$ are the atoms of $\mathcal{A}^*$.

In Chapter 2 and Chapter 3 we used generally the notation Γ (instead of G_Γ) for the subsets of M.

Clearly the *cube elements* of $\mathcal{A}^*$ are the cubes defined by Definition 2.1.1, etc. Representations of elements of $\mathcal{A}^*$ by implicants (implicates) now are representations of subsets of M by unions of cubes (intersections of anticubes). The definition of prime implicants and prime implicates is obvious.

2. Subsets of K

From 1. with $a_{i\kappa} = \kappa$ for $i \in N_n, \kappa \in \{0, ..., k_i\}$ we obtain

$$\Omega_i = M_i = \{0, ..., k_i\}, i \in N_n,$$
$$A'_{i\kappa} = \{\kappa\}, \kappa \in \{0, ..., k_i\},$$

$$\Omega = K,$$
$$A_{i\kappa} = \{(x_1, ..., x_n) \in K : x_i = \kappa\}, i \in N_n, \kappa \in \{0, ..., k_i\},$$
$$A_j = \bigcap_{i \in N_n} A_{ij_i} = \{(j_1, ..., j_n)\} = \{j\} \text{ with } j \in K.$$
$$A_\Gamma = \bigcup_{j \in \Gamma} A_j = \bigcup_{j \in \Gamma} \{j\} = \Gamma.$$

The *cube elements* are cubes of the form

$$\bigtimes_1^n K_i$$

with $\emptyset \neq \times K_i \subset K$, etc.

3. Binary functions on M

The binary functions treated in Chapter 2 and 3 are the indicators 1_{A_Γ} with $\Gamma \subseteq K$ corresponding to 1. The system $\mathcal{I}^* := \{1_{A_\Gamma} : \Gamma \subseteq K\}$ is the indicator algebra generated by the indicators $1_{A_{i\kappa}}$ with

$$1_{A_{i\kappa}}(x) = \begin{cases} 1 & \text{if } x_i = a_{i\kappa} \\ 0 & \text{otherwise} \end{cases}$$

($i \in N_n$, $\kappa \in \{0, ..., k_i\}$) (see Section 8.2).

4. Binary functions on K

In case $M = K$ we have $A_\Gamma = \Gamma$ and so $1_{A_\Gamma} = 1_\Gamma$,$\Gamma \subseteq K$. Now the system $\mathcal{I}^* := \{1_\Gamma : \Gamma \subseteq K\}$ is the indicator algebra generated by the indicators $1_{A_{i\kappa}}$ with

$$1_{A_{i\kappa}}(x) = \begin{cases} 1 & \text{if } x_i = \kappa \\ 0 & \text{otherwise} \end{cases}$$

($i \in N_n$, $\kappa \in \{0, ..., k_i\}$) (see Section 8.2).

8.2 Indicator Algebras

Indicator functions (or shortly indicators) are a useful tool in set theory and especially in probability theory. Binary functions are special indicators defined on a finite Cartesian product $\times M_i$.

In the following let Ω be always a nonempty set. We consider indicators defined for all subsets of Ω.

8.2.1 Definition

For $A \subseteq \Omega$, the *indicator* $1_A : \Omega \to \{0,1\}$ is defined by

$$1_A(\omega) := \begin{cases} 1 & \text{for } \omega \in A \\ 0 & \text{for } \omega \in \overline{A}. \end{cases}$$

First we will see, that the set of all indicators 1_A with $A \in \mathcal{A} \subseteq \mathcal{P}(\Omega)$ is a Boolean algebra if and only if $\mathcal{A}$ is an algebra in Ω.

8.2.2 Theorem

Let $\mathcal{A} \subseteq \mathcal{P}(\Omega)$ and $\mathcal{I} := \{1_A : A \in \mathcal{A}\}$.

Then the following are equivalent:

(a) $\mathcal{A}$ is an algebra in Ω.

(b) $\mathcal{I}$ is a Boolean algebra with the usual order relation $\leq$ given by

$$1_A \leq 1_{A'} :\Leftrightarrow A \subseteq A'$$

and

$$\begin{aligned}
&\mathbb{O} = 1_\phi \equiv 0, \ \mathbb{1} = 1_\Omega \equiv 1, \\
&1_A \vee 1_{A'} = \max(1_A, 1_{A'}) = 1 - (1 - 1_A)(1 - 1_{A'}) = 1_{A \cup A'}, \\
&1_A \wedge 1_{A'} = \min(1_A, 1_{A'}) = 1_A \cdot 1_{A'} = 1_{A \cap A'}, \\
&\neg 1_A = 1 - 1_A = 1_{\overline{A}}.
\end{aligned}$$

We may express this by the "modified truth table"

1_A	$1_{A'}$	$1_A \vee 1_{A'}$	$1_A \wedge 1_{A'}$	$\neg 1_A$
0	0	0	0	1
0	1	1	0	1
1	0	1	0	0
1	1	1	1	0

Obviously $1_A \leq 1_{A'}$ holds if and only if for each $\omega \in \Omega$ either $1_A(\omega) = 0$ or $1_A(\omega) = 1_{A'}(\omega) = 1$.

Proof

The proposition follows from Theorem 7.1.5 by the well–known indicator properties.

8.2.3 Theorem

Let $\{A_{10},...,A_{1k_1}\},...,\{A_{n0},...,A_{nk_n}\}$ be n partitions of Ω.
Then the indicator system

$$\mathcal{I}^* := \{1_{A_\Gamma} : \Gamma \subseteq K\} = \{1_A : A \in \mathcal{A}^*\}$$

is a Boolean algebra with the order relation $1_{A_\Gamma} \leq 1_{A_{\Gamma'}} \Leftrightarrow A_\Gamma \subseteq A_{\Gamma'}$ and

$$\begin{aligned}
&\mathcal{O} = 1_{A_\phi} = 1_\phi \equiv 0,\\
&\mathbb{1} = 1_{A_K} = 1_\Omega \equiv 1,\\
&1_{A_\Gamma} \vee 1_{A_{\Gamma'}} = \max(1_{A_\Gamma}, 1_{A_{\Gamma'}}) = 1 - (1 - 1_{A_\Gamma})(1 - 1_{A_{\Gamma'}}) = 1_{A_{\Gamma\cup\Gamma'}},\\
&1_{A_\Gamma} \wedge 1_{A_{\Gamma'}} = \min(1_{A_\Gamma}, 1_{A_{\Gamma'}}) = 1_{A_\Gamma} \cdot 1_{A_{\Gamma'}} = 1_{A_{\Gamma\cap\Gamma'}},\\
&\neg 1_{A_\Gamma} = 1 - 1_{A_\Gamma} = 1_{A_{\overline{\Gamma}}}.
\end{aligned}$$

The elements 1_{A_Γ} of $\mathcal{I}^*$ are given by

$$(8.2.4) \qquad 1_{A_\Gamma} = \sum_{j\in\Gamma} \prod_{i\in N_n} 1_{A_{ij_i}} \quad \text{for } \Gamma \subseteq K,\ 1_{A_\emptyset} := \mathcal{O} \equiv 0\,.$$

The elements $1_{A_j} := \prod_{i\in N_n} 1_{A_{ij_i}}$, for $j := (j_1, ..., j_n) \in K$, (see Theorem 8.1.1) with $1_{A_j} \neq \mathcal{O}$ are the atoms of $\mathcal{I}^*$.

Further $\{1_{A_{10}}, ..., 1_{A_{1k_1}}\}, ..., \{1_{A_{n0}}, ..., 1_{A_{nk_n}}\}$ are partitions of $\mathbb{1}$. This is equivalent to

$$\sum_{k=0}^{k_i} 1_{A_{ik}} = 1 \quad \text{for } i \in N_n\,.$$

The system $\mathcal{I}^*$ is the Boolean algebra generated by these n partitions of $\mathbb{1}$.

Proof

From Theorem 8.1.1 and Theorem 8.2.2 we conclude that $\mathcal{I}^*$ is a Boolean algebra with the proposed properties. From

$$A_\Gamma = \bigcup_{j\in\Gamma} \bigcap_{i\in N_n} A_{ij_i} = \bigcup_{j\in\Gamma} A_j$$

we obtain (8.2.4), since the sets $A_j, j \in K$ are mutually disjoint.
By $A_{i0} \cup ... \cup A_{ik_i} = \Omega,\ A_{ik} \cap A_{ik'} = \emptyset$, for $k \neq k', i \in N_n$ we have

$$\bigvee_{k=0}^{k_i} 1_{A_{ik}} := \max_{0\leq k\leq k_i} 1_{A_{ik}} = 1, 1_{A_{ik}} \cdot 1_{A_{ik'}} = 0 \quad \text{for } k \neq k'\,,$$

which is equivalent to

$$\sum_{k=0}^{k_i} 1_{A_{ik}} = 1 .$$

Thus $\{1_{A_{10}}, ..., 1_{A_{1k_1}}\}, ..., \{1_{A_{n0}}, ..., 1_{A_{nk_n}}\}$ are n partitions of $\mathbb{1}$. Now Theorem 7.2.3 and (8.2.4) yields that $\mathcal{I}^*$ is the Boolean algebra generated by these n partitions of $\mathbb{1}$ with the atoms $1_{A_i} \neq \mathcal{O}$. □

We note that $\{1_{A_{10}}, ..., 1_{A_{nk_n}}\} \subseteq \mathcal{I}^*$ (see the proof of Theorem 7.2.3).

8.2.5 Properties of $\mathcal{I}^* = \{1_{A_\Gamma}(\omega) : \Gamma \subseteq K\} = \{1_A(\omega) : A \in \mathcal{A}^*\}$

Suppose $1_{A_j} \neq \mathcal{O}$ for all $j \in K$. Then again the list 7.2.5 may be read as a list of properties of $\mathcal{I}^*$. We only have to replace B_j by 1_{A_j}, $\mathcal{O}$ by $1_\emptyset = 0$, $\mathbb{1}$ by $1_\Omega = 1$, $B_\Gamma \vee B_{\Gamma'}$ by $\max(1_{A_\Gamma}, 1_{A_{\Gamma'}})$, $B_\Gamma \wedge B_{\Gamma'}$ by $\min(1_{A_\Gamma}, 1_{A_{\Gamma'}}) = 1_{A_\Gamma} \cdot 1_{A_{\Gamma'}}, \neg B_\Gamma$ by $1 - 1_{A_\Gamma}$.

8.2.6 Representations of Indicators out of $\mathcal{I}^*$ by Implicants

Now we will state representations of indicators $1_{A_\Gamma} \in \mathcal{I}^*$ by implicants.

8.2.6.1 Definition

For $\phi \neq \times K_i \subset K$,

(8.2.6.2) $$C_I(K_1, ..., K_n) := \prod_{i \in N_n} \sum_{\kappa \in K_i} 1_{A_{ik}} = 1_{A_{\times K_i}} \in \mathcal{I}^*$$

is called a *cube element* of $\mathcal{I}^*$.

Remark

Obviously $C_I(K_1, ..., K_n)$ is the indicator of the set cube element $A_{\times K_i}$ of $\mathcal{A}^*$ given by (8.1.3.2). If $M = K$ and $1_{A_j} \neq 0$ (i.e. $A_j \neq \emptyset$) for each $j \in K$, then 1_{A_Γ} is a cube element if and only if Γ is a cube according to Definition 2.1.1.

8.2.6.3 Definition

A cube element C_I of $\mathcal{I}^*$ is called an *implicant* of 1_{A_Γ} if $C_I \leq 1_{A_\Gamma}$. An implicant C_I of 1_{A_Γ} is called a *prime implicant* of 1_{A_Γ} if there is no implicant C'_I of 1_{A_Γ} with $C_I < C'_I$ (i.e. $C_I \leq C'_I, C_I \neq C'_I$).

8.2.6.4 Definition

Let $C_{I1}, ..., C_{Ir}$ be implicants (prime implicants) of 1_{A_Γ} with

(8.2.6.5) $$1_{A_\Gamma} = \max_{\rho \in N_r} C_{I\rho} .$$

Then (8.2.6.5) is called a *representation of* 1_{A_Γ} *by implicants (prime implicants).* As to the obvious definition of *reduced* representations, *minimal* representations and *essential* prime implicants we refer again to the general Definitions 7.3.8, 7.3.9, 7.3.10, (see also the Definitions 8.1.3.6, 8.1.3.7, 8.1.3.8).

Now we state the indicator version of Theorem 7.3.11.

8.2.6.6 Theorem

Suppose $A_j \neq \emptyset$, for each $j \in K$. Let $M = K$. Then, for $\Gamma \subset K$, the following hold.

(a) A cube element $C_I(K_1, ..., K_n)$ of $\mathcal{I}^*$ is an implicant (prime implicant, essential prime implicant) of $1_{A_\Gamma} \in \mathcal{I}^*$ if and only if the cube indicator $C(K_1, ..., K_n)$ (Definition 2.1.1) is an implicant (prime implicant, essential prime implicant) of 1_Γ.

(b) The indicator $1_{A_\Gamma} \in \mathcal{I}^*$ has the representation (reduced representation, minimal representation (8.2.6.5)) by the implicants (prime implicants) $C_{I1}, ..., C_{Ir}$ with $C_{I\rho} := C_I(K_{\rho 1}, ..., K_{\rho n})$ (conf. also (8.2.6.2)) for $\rho \in N_r$ if and only if the indicator 1_Γ has the representation (reduced representation, minimal representation)

$$1_\Gamma = \max_{\rho \in N_r} C_\rho$$

by the implicants (prime implicants) $C_1, ..., C_r$ with $C_\rho := C(K_{\rho 1}, ..., K_{\rho n})$ for $\rho \in N_r$.

8.2.7 Representation of Indicators out of $\mathcal{I}^*$ by Implicates

In the following we state again the implicates counterpart of the representations by implicants.

8.2.7.1 Definition

Let $\phi \neq \times K_i \subset K$. Then the element $F_I(K_1, ..., K_n)$ of $\mathcal{I}^*$ given by

$$F_I(K_1, ..., K_n) := \max_{i \in N_n} \sum_{k \in K_i} 1_{A_{ik}} \; (= \max_{i \in N_n} \prod_{k \in \overline{K_i}} (1 - 1_{A_{ik}}))$$

is called an *anticube element* of $\mathcal{I}^*$.

Remark

It holds

$$F_I(K_1, \ldots, K_n) = 1 - C_I(\overline{K}_1, \ldots, \overline{K}_n) = 1_{A_{K \setminus \times \overline{K}_i}} \,. \tag{8.2.7.3}$$

If $M = K$ and $1_{A_j} \neq 0$ (i.e. $A_j \neq \emptyset$) for each $j \in K$, then 1_{A_Γ} is an anticube element if and only if Γ is an anticube according to Definition 3.1.1.

8.2.7.4 Definition

An anticube element F_I of $\mathcal{I}^*$ is called an *implicate* of 1_{A_Γ} if $1_{A_\Gamma} \leq F_I$. An implicant F_I of 1_{A_Γ} is called a *prime implicate* of 1_{A_Γ} if there is no implicate F_I' of 1_{A_Γ} with $F_I' < F_I$.

8.2.7.5 Definition

Let $F_{I1}, \ldots, F_{I_r}$ be implicates (prime implicates) of 1_{A_Γ} with

$$1_{A_\Gamma} = \min_{\rho \in N_r} F_{I_\rho} = \prod_{\rho \in N_r} F_{I_\rho} \,. \tag{8.2.7.6}$$

Then (8.2.7.6) is called *a representation of* 1_{A_Γ} *by implicates (prime implicates).*

We omit the obvious definition of reduced (minimal) representations and essential prime implicates.

The next is the indicator version of Theorem 7.4.11.

8.2.7.7 Theorem

Suppose $A_j \neq \emptyset$, for each $j \in K$. Let $M = K$. Then, for $\Gamma \subset K$ the following hold.

(a) An anticube element $F_I(K_1, \ldots, K_n)$ of $\mathcal{I}^*$ is an implicate (prime implicate, essential prime implicate) of $1_{A_\Gamma} \in \mathcal{I}^*$ if and only if the anticube indicator $F(K_1, \ldots, K_n)$ (Definition 3.1.1) is an implicate (prime implicate, essential prime implicate) of 1_Γ.

(b) The indicator $1_{A_\Gamma} \in \mathcal{I}^*$ has the representation (reduced representation, minimal representation)

$$1_{A_\Gamma} = \min_{\rho \in N_r} F_{I_\rho} = \prod_{\rho \in N_r} F_{I_\rho}$$

by the implicates (prime implicates) $F_{I1}, ..., F_{Ir}$ with $F_{I\rho} := F_I(K_{\rho 1}, ..., K_{\rho n})$ for $\rho \in N_r$ if and only if 1_Γ has the representation (reduced representation, minimal representation)

$$1_\Gamma = \min_{\rho \in N_r} F_\rho = \prod_{\rho \in N_r} F_\rho$$

by the implicates (prime implicates) $F_1, ..., F_r$ with $F_\rho := F(K_{\rho 1}, ..., K_{\rho n})$ for $\rho \in N_r$.

8.3 Partitions in Propositional Logic

The following considerations will be the basis for Boolean algebras of classes of equivalent logical propositions. These finitely generated Algebras will be treated in Section 8.4.

In propositional logic the propositions are introduced as indefined things together with some logical connectives. We call such a system a propositional calculus (see Hailperin (1976)).

8.3.1 Definition

A system $\mathcal{R}$ of any things – called *propositions* – is called a *propositional calculus* if the following holds:

$$\text{If } p, q \in \mathcal{R} \text{ then } \begin{cases} \neg p \\ p \vee q \\ p \wedge q \\ p \rightarrow q \\ p \leftrightarrow q \end{cases} \in \mathcal{R},$$

where the logical connectives

$\neg$ (negation "not")

$\vee$ (disjunction "or")

$\wedge$ (conjunction "and")

$\rightarrow$ (subjunction "if ... then")

$\leftrightarrow$ (bijunction "if and only if ... then")

are formal operations defined on $\mathcal{R}$. (Note that $\leftrightarrow$ does not mean equivalent.)

We consider propositions as statements concerning certain events or facts in such a way that a proposition may be true, if the appropriate event occurs, and false otherwise. To state the truth behaviour of all elements of a propositional calculus we give the following usual definition.

8.3.2 Definition

Let $p, q \in \mathcal{R}$. Then

$\neg p$ is true if p is false.
$p \vee q$ is true if p or q (or both) are true.
$p \wedge q$ is true if p and q are true.
$p \rightarrow q$ is true if p is false or p and q are true.
$p \leftrightarrow q$ is true if p and q are true or p and q are false.

Remark

For the practical determination of the truth behaviour of propositions it is suitable to use the following well-known truth-table.

p	q	$\neg p$	$p \vee q$	$p \wedge q$	$p \rightarrow q$	$p \leftrightarrow q$
F	F	T	F	F	T	T
F	T	T	T	F	T	F
T	F	F	T	F	F	F
T	T	F	T	T	T	T.

Moreover we will see later that indicators are a more convenient tool to treat complicate logical expressions.

8.3.3 Definition

Let $p_1, ..., p_r$ be propositions and $A(p_1, ..., p_r), B(p_1, ..., p_r)$ propositions formed by logical connectives from $p_1, ..., p_r$. Then $A(p_1, ..., p_r)$ and $B(p_1, ..., p_r)$ are called *equivalent* – written $A(p_1, ..., p_r) \sim B(p_1, ..., p_r)$ – if both are true or both are false, for all truth values of $p_1, ..., p_r$.

If $A(p_1, ..., p_r)$ is always true, we write $A(p_1, ..., p_r) \sim \mathbb{T}$, if $A(p_1, ..., p_r)$ is always false, we write $A(p_1, ..., p_r) \sim \mathbb{F}$, where $\mathbb{T}$ is a proposition which is always true, and $\mathbb{F}$ is a proposition which is always false. Obviously $A(p_1, ..., p_r) \sim B(p_1, ..., p_r)$ is the same as $A(p_1, ..., p_r) \leftrightarrow B(p_1, ..., p_r) \sim \mathbb{T}$.

Examples

$p_1 \wedge (p_2 \vee p_3) \sim (p_1 \wedge p_2) \vee (p_1 \wedge p_3)$,
$p_1 \rightarrow p_2 \sim \neg p_1 \vee p_2$,
$p_1 \vee \neg p_1 \sim \mathbb{T}$,
$p_1 \wedge \neg p_1 \sim \mathbb{F}$.

8.3.4 Theorem

Let $p_1, ..., p_r$ be any propositions and $\mathcal{R}(\{p_1, ..., p_r\})$ the system of all propositions formed stepwise from $p_1, ..., p_r$ by logical connectives. Then $\mathcal{R}(\{p_1, ..., p_r\})$ is the smallest propositional calculus containing $\{p_1, ..., p_r\}$. We call it the propositional calculus *generated* by $\{p_1, ..., p_r\}$.

Proof

The assertion is obvious.

Obviously $\mathcal{R}(\{p_1, ..., p_r\})$ contains countably infinitely many elements.

Remark

The relation $\sim$ is an equivalence relation on $\mathcal{R}(p_1, ..., p_r)$, i.e. for $p, q, r \in \mathcal{R}(p_1, ..., p_r)$ holds

$$\begin{aligned} &p \sim p \\ &p \sim q \text{ implies } q \sim p \\ &p \sim q, q \sim r \text{ implies } p \sim r. \end{aligned} \tag{8.3.5}$$

The propositional calculus $\mathcal{R}(\{p_1, ..., p_r\})$ is not a Boolean algebra. E.g. the propositions $p_1 \wedge (p_2 \vee p_3)$ and $(p_1 \wedge p_2) \vee (p_1 \wedge p_3)$ are considered as (equivalent but) different propositions. Thus the corresponding distribution law is not fulfilled.

8.3.6 Definition

Let $q_1, ..., q_m$ be propositions with

$$\begin{aligned} &q_i \not\sim \mathbb{F} \text{ for } i \in N_m, \\ &q_i \wedge q_j \sim \mathbb{F} \text{ for } i, j \in N_m, i \neq j \end{aligned}$$

and

$$\bigvee_{i \in N_m} q_i \sim \mathbb{T}.$$

Then we call $\{q_1, ..., q_m\}$ a *partition* of $\mathbb{T}$.

In a partition of $\mathbb{T}$ exactly one of the propositions $q_1, ..., q_r$ (we suppose $r > 1$) is true. No q_i is always false, no q_i is always true.

8.4 Algebras of Classes of Propositions

Let us consider any propositional calculus $\mathcal{R}$. Since $p \sim q$ is an equivalence relation (see (8.3.5)) we obtain a partition of $\mathcal{R}$ in mutually disjoint equivalence classes p^* where p^* contains all propositions which are equivalent to p.

First we show that the system of all equivalence classes of $\mathcal{R}$ is again a propositional calculus.

8.4.1 Theorem

Let $\mathcal{R}$ be a propositional calculus according to Definition 8.3.1 and Definition 8.3.2 with the equivalence relation given by Definition 8.3.3. Let p^* be the class of all propositions which are equivalent to $p \in \mathcal{R}$. Then the system

$$\Phi(\mathcal{R}) := \{p^* : p \in \mathcal{R}\}$$

of all classes of equivalent propositions is a propositional calculus with

$$\begin{aligned} \neg p^* &:= (\neg p)^*, \\ p^* \vee q^* &:= (p \vee q)^*, \\ p^* \wedge q^* &:= (p \wedge q)^*, \\ p^* \rightarrow q^* &:= (p \rightarrow q)^*, \\ p^* \leftrightarrow q^* &:= (p \leftrightarrow q)^*. \end{aligned}$$

(Note that $p^* = q^*$ if and only if $p \sim q$.)

Proof

We have to prove:

$$\text{If } p^*, q^* \in \Phi(\mathcal{R}) \text{ then } \begin{cases} \neg p^* \\ p^* \vee q^* \\ p^* \wedge q^* \\ p^* \rightarrow q^* \\ p^* \leftrightarrow q^* \end{cases} \in \quad \Phi(\mathcal{R}).$$

Suppose $p^* \in \Phi(\mathcal{R})$. Then $p \in \mathcal{R}$ and so $\neg p \in \mathcal{R}$ implying $(\neg p^*) \in \Phi(\mathcal{R})$. Suppose $p^*, q^* \in \Phi(\mathcal{R})$. Then $p, q \in \mathcal{R}$ and so $p \vee q \in \mathcal{R}$ implying $(p \vee q)^* \in \Phi(\mathcal{R})$. In the same way the remaining assertions are proved. □

Next we show that $\Phi(\mathcal{R})$ is a Boolean algebra.

8.4.2 Theorem

Let $\Phi(\mathcal{R}) := \{p^* : p \in \mathcal{R}\}$ given as in Theorem 8.4.1. Then $\Phi(\mathcal{R})$ is a Boolean algebra with the order relation $\leq$ defined by

$$p^* \leq q^* :\Leftrightarrow (p \rightarrow q)^* = \mathbb{T}^* \quad \text{(i.e. } p \rightarrow q \sim \mathbb{T}\text{)}$$

and

$$\mathbb{0} = \mathbb{F}^* \ , \ \mathbb{1} = \mathbb{T}^*.$$

Proof

From $p \rightarrow p \sim \mathbb{T}$ it follows $p^* \leq p^*$. Now suppose $p^* \leq q^*$ and $q^* \leq p^*$. Then $p \rightarrow q \sim \mathbb{T}$ and $q \rightarrow p \sim \mathbb{T}$, thus $p \leftrightarrow q \sim \mathbb{T}$ (due to Definition 8.3.2) and so $p \sim q$ implying $p^* = q^*$. Finally suppose $p^* \leq q^*$ and $q^* \leq r^*$. Then $p \rightarrow q \sim \mathbb{T}$ and $q \rightarrow r \sim \mathbb{T}$. Then $p \rightarrow r \sim \mathbb{T}$ (due to Definition 8.3.2) and so $p^* \leq r^*$. Thus $\leq$ is an order relation.

From $p \rightarrow p \vee q \sim \mathbb{T}$ and $q \rightarrow p \vee q \sim \mathbb{T}$ we obtain $p^* \leq (p \vee q)^* = p^* \vee q^*$ and $q^* \leq (p \vee q)^* = p^* \vee q^*$. Now suppose $p^* \leq r^*$ and $q^* \leq r^*$. Then $p \rightarrow r \sim \mathbb{T}$ and $q \rightarrow r \sim \mathbb{T}$. This implies $p \vee q \rightarrow r \sim \mathbb{T}$ and so $p^* \vee q^* = (p \vee q)^* \leq r^*$. Thus $p^* \vee q^* = \sup(p^*, q^*)$. In the same way we show $p^* \wedge q^* = \inf(p^*, q^*)$. By definition we have

$$\begin{aligned} p^* \wedge (q^* \vee r^*) &= p^* \wedge (q \vee r)^* = (p \wedge (q \vee r))^* = ((p \wedge q) \vee (p \wedge r))^* \\ &= (p \wedge q)^* \vee (p \wedge r)^* = (p^* \wedge q^*) \vee (p^* \wedge r^*). \end{aligned}$$

The other distribution law is proved in the same way.

From $p \rightarrow \mathbb{T} \sim \mathbb{T}$ and $\mathbb{F} \rightarrow p \sim \mathbb{T}$ finally we obtain $\mathbb{F}^* \leq p^* \leq \mathbb{T}^*$ for all $p \in \Phi(\mathcal{R})$.

Clearly $F^* = (p \wedge (\neg p))^* \in \Phi(\mathcal{R})$ and $\mathbb{T}^* = (p \vee (\neg p))^* \in \Phi(\mathcal{R})$. □

Remark

From $p \rightarrow q \sim (\neg p) \vee q$ and $p \leftrightarrow q \sim (p \wedge q) \vee ((\neg p) \wedge (\neg q))$ we obtain

$$\begin{aligned} p^* \rightarrow q^* &:= (p \rightarrow q)^* = (\neg p^*) \vee q^*, \\ p^* \leftrightarrow q^* &:= (p \leftrightarrow q)^* = (p^* \wedge q^*) \vee ((\neg p^*) \wedge (\neg q^*)) \end{aligned}$$

so that the operations $\rightarrow$ and $\leftrightarrow$ are defined also in the Boolean algebra $\Phi(\mathcal{R})$ by the Boolean operations $\neg, \vee, \wedge$.

Now let us consider Boolean algebras (of classes of equivalent propositions) which are generated by n partitions of $\mathbb{T}^*$.

8.4.3 Definition

Let $p_1^*, ..., p_r^* \in \Phi(\mathcal{R})$ with

$$p_i^* \neq \mathbb{F}^* \quad \text{for} \quad i \in N_r,$$
$$p_i^* \wedge p_j^* = \mathbb{F}^* \quad \text{for} \quad i, j \in N_r, i \neq j$$

and

$$\bigvee_{i \in N_r} p_i^* = \mathbb{T}^*.$$

Then we call $\{p_1^*, ..., p_r^*\}$ a *partition of* $\mathbb{T}^*$.

We note that if $p_1, ..., p_r$ is a partition of $\mathbb{T}$ according to Definition 8.3.6 then $\{p_1^*, ..., p_r^*\}$ is a partition of $\mathbb{T}^*$. Conversely if $\{p_1^*, ..., p_r^*\}$ is a partition of $\mathbb{T}^*$ then $\{q_1, ..., q_r\}$ is a partition of $\mathbb{T}$ for all $q_1, ..., q_r$ with $q_1 \sim p_1, ..., q_r \sim p_r$.

8.4.4 Theorem

Let $\mathcal{R}$ be a propositional calculus as in Theorem 8.4.1 and $\{p_{10}, ..., p_{1k_1}\}, ..., \{p_{n0}, ..., p_{nk_n}\}$ n partitions of $\mathbb{T}$. Then $\{p_{10}^*, ..., p_{1k_1}^*\}, ..., \{p_{n0}^*, ..., p_{nk_n}^*\} \subseteq \Phi(\mathcal{R})$ are n partitions of $\mathbb{T}^*$, and the Boolean algebra (of classes of propositions) $\Phi^* := \Phi(\{p_{10}^*, ..., p_{nk_n}^*\}) \subseteq \Phi(\mathcal{R})$ generated by $\{p_{10}^*, ..., p_{nk_n}^*\}$ is given by

$$\Phi^* = \{p_\Gamma^* : \Gamma \subseteq K\}$$

with

$$p_\Gamma := \bigvee_{j \in \Gamma} \bigwedge_{i \in N_n} p_{ij_i}$$

and so

$$p_\Gamma^* = (\bigvee_{j \in \Gamma} \bigwedge_{i \in N_n} p_{ij_i})^* = \bigvee_{j \in \Gamma} \bigwedge_{i \in N_n} p_{ij_i}^* \quad \text{for } \Gamma \subseteq K, p_\emptyset^* := \emptyset = \mathbb{F}^*.$$

It holds

$$p_K^* = \mathbb{1} = \mathbb{T}^*, p_\Gamma^* \vee p_{\Gamma'}^* = p_{\Gamma\cup\Gamma'}^*, p_\Gamma^* \wedge p_{\Gamma'}^* = p_{\Gamma\cap\Gamma'}^*, \neg p_\Gamma^* = p_{\overline{\Gamma}}^*,$$
$$p_\Gamma^* \leq p_{\Gamma'}^* \Leftrightarrow p_{\overline{\Gamma}\cup\Gamma'}^* = \mathbb{1}.$$

The classes $p_j^* := p_{\{j\}}^* = (\bigwedge_{i\in N_n} p_{ij_i})^* = \bigwedge_{i\in N_n} p_{ij_i}^*$ with $p_j^* \neq \mathbb{O}$ are the atoms of Φ^*. They form a partition of $\mathbb{T}$:

$$p_i^* \wedge p_k^* = \mathbb{O} \text{ for } j \neq k, \bigvee_{j\in K} p_j^* = \mathbb{1} = \mathbb{T}^*.$$

Let $\mathcal{R}(\{p_{10}, ..., p_{nk_n}\})$ be the propositional calculus generated by $\{p_{10}, ..., p_{nk_n}\}$ (see Theorem 8.3.4). Then it holds also

$$\Phi^* = \{p^* : p \in \mathcal{R}(\{p_{10}, ..., p_{nk_n}\})\}.$$

This yields

$$\mathcal{R}(\{p_{10}, ..., p_{nk_n}\}) = \dot{\bigcup_{\Gamma\subseteq K}} p_\Gamma^*.$$

Proof

Obviously $\{p_{10}^*, ..., p_{1k_1}^*\}, ..., \{p_{n1}^*, ..., p_{nk_n}^*\}$ are n partitions of $\mathbb{T}^*$. Since $\Phi(\mathcal{R})$ is a Boolean algebra (with the order relation $p^* \leq q^* \Leftrightarrow p \to q \sim \mathbb{T}$), from Theorem 7.2.3 we obtain the essential statement of Theorem 8.4.4. Only

(a) $(\bigvee_{j\in\Gamma} \bigwedge_{i\in N_n} p_{ij_i})^* = \bigvee_{j\in\Gamma} \bigwedge_{i\in N_n} p_{ij_i}^*$,

(b) $p_\Gamma^* \leq p_{\Gamma'}^* \Leftrightarrow p_{\overline{\Gamma}\cup\Gamma'}^* = \mathbb{1}$

(c) $\Phi^* = \{p^* : p \in \mathcal{R}(\{p_{10}, ..., p_{nk_n}\})\}$

has to be proved.

(a): The relation follows from Theorem 8.4.1 and Theorem 8.4.2: $\Phi(\mathcal{R})$ is a Boolean algebra with the Boolean operations given in Theorem 8.4.1.

(b): From $p^* \leq q^* \Leftrightarrow (p \to q)^* = \mathbb{1}$ and $p \to q \sim (\neg p) \vee q$ we obtain $(p \to q)^* = ((\neg p) \vee q)^* = (\neg p)^* \vee q^* = (\neg p^*) \vee q^*$. With $p = p_\Gamma, q = p_{\Gamma'}$ this yields

$$p_\Gamma^* \leq p_{\Gamma'}^* \Leftrightarrow (p_\Gamma \to p_{\Gamma'})^* = \mathbb{1} \Leftrightarrow (\neg p_\Gamma^*) \vee p_{\Gamma'}^* = \mathbb{1}$$
$$\Leftrightarrow p_{\overline{\Gamma}}^* \vee p_{\Gamma'}^* = \mathbb{1} \Leftrightarrow (p_{\overline{\Gamma}\cup\Gamma'})^* = \mathbb{1}.$$

(c): Write $\Phi' := \{p^* : p \subset \mathcal{R}(\{p_{10}, ..., p_{nk_n}\})\}$. Suppose $p_\Gamma^* \in \Phi^*$. Obviously

then $p_\Gamma \in \mathcal{R}(\{p_{10}, ..., p_{nk_n}\})$ and so $p_\Gamma^* \in \Phi'$. Now suppose $p^* \in \Phi'$, i.e. $p \in \mathcal{R}(\{p_{10}, ..., p_{nk_n}\})$.
Define $\Gamma(p)$ by

$$\Gamma(p) := \{j \in K : p \text{ is true if } \bigwedge_{i \in N_n} p_{ij_i} \text{ is true}\}.$$

Then p is true if and only if $\bigwedge_{i \in N_n} p_{ij_i}$ is true for exactly one $j \in \Gamma(p)$, i.e. if and only if

$$p_{\Gamma(p)} = \bigvee_{j \in \Gamma(p)} \bigwedge_{i \in N_n} p_{ij_i}$$

is true. Thus p is equivalent to $p_{\Gamma(p)}$ with $\Gamma(p) \subseteq K$ i.e. $p_{\Gamma(p)}^* \in \Phi^*$ and so we have $p^* = p_{\Gamma(p)}^* \in \Phi^*$. □

8.4.5 Supplement

Define the set operations $\rightarrow$ and $\leftrightarrow$ on $\mathcal{P}(K)$ by

$$\Gamma \rightarrow \Gamma' := \overline{\Gamma} \cup \Gamma',$$
$$\Gamma \leftrightarrow \Gamma' := (\Gamma \cap \Gamma') \cup (\overline{\Gamma} \cap \overline{\Gamma'}).$$

Then for the operations $p_\Gamma^* \rightarrow p_{\Gamma'}^* := (p_\Gamma \rightarrow p_{\Gamma'})^*$ and $p_\Gamma^* \leftrightarrow p_{\Gamma'}^* := (p_\Gamma \leftrightarrow p_{\Gamma'}^*)$ hold

$$p_\Gamma^* \rightarrow p_{\Gamma'}^* = p_{\Gamma \rightarrow \Gamma'}^*,$$
$$p_\Gamma^* \leftrightarrow p_{\Gamma'}^* = p_{\Gamma \leftrightarrow \Gamma'}^*.$$

Proof

This follows from $p_\Gamma \rightarrow p_{\Gamma'} \sim \neg p_\Gamma \vee p_{\Gamma'}$ and $p_\Gamma \leftrightarrow p_{\Gamma'} \sim (p_\Gamma \wedge p_{\Gamma'}) \vee (\neg p_\Gamma \wedge \neg p_{\Gamma'})$ yielding

$$(p_\Gamma \rightarrow p_{\Gamma'})^* = (\neg p_\Gamma)^* \vee p_{\Gamma'}^* = p_{\overline{\Gamma}}^* \vee p_{\Gamma'}^* = p_{\overline{\Gamma} \cup \Gamma'}^* = p_{\Gamma \rightarrow \Gamma'}^*$$

and

$$(p_\Gamma \leftrightarrow p_{\Gamma'})^* = (p_\Gamma \wedge p_{\Gamma'})^* \vee (\neg p_\Gamma \wedge \neg p_{\Gamma'})^* = p_{\Gamma \cap \Gamma'}^* \vee p_{\overline{\Gamma} \cap \overline{\Gamma'}}^*$$
$$= p_{(\Gamma \cap \Gamma') \cup (\overline{\Gamma} \cap \overline{\Gamma'})}^* = p_{\Gamma \rightarrow \Gamma'}^*.$$
□

8.4.6 Properties of Φ^*

Suppose $p_j^* \; (:= p_{\{j\}}^* = \bigwedge_{i \in N_n} p_{ij_i}^*) \neq \mathbb{O} \; (\neq \mathbb{F}^*)$ for all $j \in K$. Then the list 7.2.5 may be read as a list of properties of Φ^*. We only have to replace B_{ij} by p_{ij}^*, B_j by p_j^*, B_Γ by p_Γ^*.

8.4.7 Representation of Elements of Φ^* by Implicants

Now representations of elements $p_\Gamma^* \in \Phi^*$ by implicants again may be stated by using the results of Section 7.3.

8.4.7.1 Definition

Let $\emptyset \neq \times K_i \subset K$. Then

$$(8.4.7.2) \quad C_p(K_1, ..., K_n) := \bigwedge_{i \in N_n} \bigvee_{k \in K_i} p_{ik}^* = (\bigwedge_{i \in N_n} \bigvee_{k \in K_i} p_{ik})^* = p^*_{\times K_i} \in \Phi^*$$

is called a *cube element* of Φ^*.

The meaning of $C_p(K_1, ..., K_n) = p^*_{\times K_i}$ is the following: The proposition $p_{\times K_i}$ is true if and only if for each $i \in N_n$ exactly one of the propositions p_{ik} with $k \in K_i$ is true.

8.4.7.3 Definition

A cube element C_p of Φ^* is called an *implicant* of p_Γ^* if $C_p \leq p_\Gamma^*$. An implicant C_p of p_Γ^* is called a *prime implicant* of p_Γ^* if and only if there is no implicant C_p' of p_Γ^* with $C_p < C_p'$.

8.4.7.4 Definition

Let $C_{p1}, ..., C_{pr}$ be implicants (prime implicants) of p_Γ^* with

$$(8.4.7.5) \qquad p_\Gamma^* = \bigvee_{\rho \in N_r} C_{p\rho}.$$

Then (8.4.7.5) is called a *representation of* p_Γ^* *by implicants (prime implicants).*

For the further definitions we refer again to the general definitions of Chapter 7.

We leave the interpretation of (8.4.7.5) to the reader, as well as the translation of Theorem 7.3.11 into the appropriate form for the elements of Φ^*. This version then again allows to use the indicator methods of Chapter 2 for the construction of implicant representations of the class p^*_Γ out of Φ^*.

8.4.8 Representation of Elements of Φ^* by Implicates

We only state the definitions of implicates and representations of elements of Φ^* by implicates.

8.4.8.1 Definition

Let $\phi \neq \times K_i \subset K$. Then

$$F_p(K_1, ..., K_n) := \bigvee_{i \in N_n} \bigvee_{k \in K_i} p^*_{ik} = p^*_{\overline{\times K_i}} \in \Phi^* \tag{8.4.8.2}$$

is called an *anticube element* of Φ^*.

Since

$$\bigvee_{i \in N_n} \bigvee_{k \in K_i} p^*_{ik} = (\bigvee_{i \in N_n} \bigvee_{k \in K_i} p_{ik})^*$$

this means: the proposition $p_{\overline{\times K_i}}$ is true if and only if for at least one $i \in N_n$ exactly one of the propositions p_{ik} with $k \in K_i$ is true.

8.4.8.2 Definition

An anticube element F_p of Φ^* is called an *implicate* of p^*_Γ if $p^*_\Gamma \leq F_p$. An implicate of p^*_Γ is called a *prime implicate* of p^*_Γ if there is no implicate F'_p of p^*_Γ with $F'_p < F_p$.

8.4.8.3 Definition

Let $F_{p1}, ..., F_{pr}$ be implicates (prime implicates) of p^*_Γ with

$$p^*_\Gamma = \bigwedge_{\rho \in N_r} F_{p\rho}. \tag{8.4.8.4}$$

Then (8.4.8.4) is called a *representation of* p^*_Γ *by implicates (prime implicates).*

Now the further definitions (reduced representations etc.) are obvious (see Section 7.4), and we may use the Φ^*–version of Theorem 7.4.11 to construct implicate respresentations of p^*_Γ.

8.5 Truth Function Algebras

In proposition logic the truth behaviour of propositions are treated by truth functions taking the values F (false) and T (true) and fulfilling certain calculating rules. We will show that systems of truth functions generated by partitions of the truth function $t \equiv T$ are again Boolean algebras of type $\mathcal{B}^*$ defined in Chapter 7.

8.5.1 Definition

Let $\mathcal{R}$ be a propositional calculus (Definition 8.3.1). Then the *coordinated truth function system* $\mathcal{T}(\mathcal{R})$ is defined by

$$\mathcal{T}(\mathcal{R}) := \{t(p) : p \in \mathcal{R}\}$$

with

$$t(p) := \begin{cases} T & \text{if } p \text{ is true} \\ F & \text{if } p \text{ is false .} \end{cases}$$

8.5.2 Theorem

Let $\mathcal{R}$ be a propositional calculus. Then $\mathcal{T}(\mathcal{R})$ is a propositional calculus with

$$\neg t(p) := t(\neg p) = \begin{cases} T & \text{if } t(p) = F \\ F & \text{if } t(p) = T, \end{cases}$$

$$t(p) \vee t(q) := t(p \vee q) = \begin{cases} F & \text{if } t(p) = t(q) = F \\ T & \text{otherwise,} \end{cases}$$

$$t(p) \wedge t(q) := t(p \wedge q) = \begin{cases} T & \text{if } t(p) = t(q) = T \\ F & \text{otherwise,} \end{cases}$$

$$t(p) \rightarrow t(q) := t(p \rightarrow q) = \begin{cases} F & \text{if } t(p) = T \text{ and } t(q) = F \\ T & \text{otherwise,} \end{cases}$$

$$t(p) \leftrightarrow (q) := t(p \leftrightarrow q) = \begin{cases} T & \text{if } t(p) = t(q) = T \text{ or } t(p) = t(q) = F \\ F & \text{otherwise.} \end{cases}$$

Proof

Suppose $t(p) \in \mathcal{T}(\mathcal{R})$. Then $p \in \mathcal{R}$ and thus $\neg p \in \mathcal{R}$ implying $t(\neg p) \in \mathcal{T}(\mathcal{R})$. Further $t(\neg p)$ equals T if $\neg p$ is true, i.e. if p is false, equivalent to $t(p) = F$, and $t(\neg p)$ equals F if $\neg p$ is false, i.e. if p is true, equivalent to $t(p) = T$.

Suppose $t(p), t(q) \in \mathcal{T}(\mathcal{R})$. Then $p, q \in \mathcal{R}$ and thus $p \vee q \in \mathcal{R}$ implying $t(p \vee q) \in \mathcal{T}(\mathcal{R})$. Further $t(p \vee q)$ equals F if and only if $p \vee q$ is false, i.e. if p is false and q is false, equivalent to $t(p) = t(q) = F$.

In the same way the other assertions are proved.

8.5.3 Theorem

Let $\mathcal{R}$ be a propositional calculus and $T(\mathcal{R})$ given by Definition 8.5.1. Then $T(\mathcal{R})$ is a Boolean algebra with the order relation $\leq$ defined by

$$t(p) \leq t(q) :\Leftrightarrow \Big(t(p) = F \;\text{ or }\; t(p) = t(q) = T\Big) \;\Leftrightarrow\; \Big(t(p) \to t(q) \equiv T\Big).$$

The connectives $\neg, \vee, \wedge$ are given according to Theorem 8.5.2. The 0–element and the 1–element are given by $\mathbb{0} :\equiv F$ and $\mathbb{1} :\equiv T$.

Proof

1. By definition it holds $t(p) \leq t(p)$. Suppose $t(p) \leq t(q)$ and $t(q) \leq t(p)$. Then $t(p) = T$ implies $t(q) = T$ due to $t(p) \leq t(q)$, and $t(p) = F$ implies $t(q) = F$ due to $t(q) \leq t(p)$. Thus $t(p) = t(q)$.
Suppose $t(p) \leq t(q)$ and $t(q) \leq t(r)$. Assume $t(p) = F$. Then always $t(p) \leq t(r)$. Assume $t(p) = T$. Then $t(q) = T$ and so $t(r) = T$, thus $t(p) \leq t(r)$. Therefore $t(p) \leq t(q)$ is an order relation.

2. From the definition it follows $t(p) \wedge \neg t(p) \equiv F$ and $t(p) \vee \neg t(p) \equiv T$. Thus $\neg t(p)$ is the complement of $t(p)$. Further obviously $F \leq t(p) \leq T$.

3. We show that $t(p) \vee t(q)$ is the supremum of $t(p)$ and $t(q)$. Suppose $t(p) = T$. Then $t(p) \vee t(q) = T$ and thus $t(p) \leq t(p) \vee t(q)$. Suppose $t(p) = F$. Then evidently $t(p) \leq t(p) \vee t(q)$. In the same way it follows $t(q) \leq t(p) \vee t(q)$. Now suppose $t(p) \leq t(r)$ and $t(q) \leq t(r)$. Then $t(r) = F$ implies $t(p) = F$ and $t(q) = F$ and so $t(p) \vee t(q) = F$, thus $t(p) \vee t(q) \leq t(r)$. Finally $t(r) = T$ implies $t(p) \vee t(q) \leq t(r)$.

4. In the same way we obtain $t(p) \wedge t(q) = \inf(t(p), t(q))$.

5. To prove the distribution law $t(p) \wedge (t(q) \vee t(r)) = (t(p) \wedge t(q)) \vee (t(p) \wedge t(r))$ we first assume $t(p) = F$. Then, as it is easy to see, both sides equals F. Now assume $t(p) = T$. Then $t(p) \wedge (t(q) \vee t(r)) = t(q) \vee t(r), t(p) \wedge t(q) = t(q)$ and $t(p) \wedge t(r) = t(r)$, thus both sides are equal.

6. The other distribution law may be proved in the same way.

Now we have proved the assertion. Partitions of the 1–element $\mathbb{1} \equiv T$ are defined by the following truth function version of Definition 8.3.5.

8.5.4 Definition

Let $t(q_1), ..., t(q_m)$ be truth functions with

$$t(q_i) \not\equiv F \quad \text{for } i \in N_m,$$
$$t(q_i) \wedge t(q_j) \equiv F \quad \text{for } i, j \in N_m, i \neq j$$

and

$$\bigvee_{i \in N_m} t(q_i) \equiv T.$$

Then we call $\{t(q_1), ..., t(q_m)\}$ a *partition* of T.

Remark

Obviously $\{t(q_1), ..., t(q_m)\}$ is a partition of T if and only if $\{q_1, ..., q_m\}$ is a partition of $\mathbb{T}$.

Now we may state the Boolean truth function algebra generated by n partitions of T. The following theorem is the complete truth function analogue of Theorem 8.4.4 and supplement 8.4.5.

8.5.5 Theorem

Let $\mathcal{R}$ be a propositional calculus according to the Definitions 8.3.1, 8.3.2 and 8.3.3, further $\{p_{10}, ..., p_{1k_1}\}, ..., \{p_{n0}, ..., p_{nk_n}\} \subseteq \mathcal{R}$ n partitions of $\mathbb{T}$.
Then $\{t(p_{10}), ..., t(p_{1k_1})\}, ..., \{t(p_{n0}), ..., t(p_{nk_n})\} \subseteq \mathcal{T}(\mathcal{R})$ are n partitions of T, and the Boolean truth function algebra $\mathcal{T}^* := \mathcal{T}(\{t(p_{10}), ..., t(p_{nk_n})\} \subseteq \mathcal{T}(\mathcal{R})$ generated by $\{t(p_{10}), ..., t(p_{nk_n})\}$ is given by

$$\mathcal{T}^* = \{t(p_\Gamma) : \Gamma \subseteq K\}$$

with (conf. Theorem 8.4.4)

$$p_\Gamma := \bigvee_{j \in \Gamma} \bigwedge_{i \in N_n} p_{ij_i}$$

and so

$$t(p_\Gamma) = t(\bigvee_{j \in \Gamma} \bigwedge_{i \in N_n} p_{ij_i}) = \bigvee_{j \in \Gamma} \bigwedge_{i \in N_n} t(p_{ij_i}) \quad \text{for } \Gamma \subseteq K, t(p_\emptyset) := \mathbb{O} \equiv F.$$

It holds
$t(p_K) = \mathbb{1} \equiv T, t(p_\Gamma) \vee t(p_{\Gamma'}) = t(p_{\Gamma \cup \Gamma'}), t(p_\Gamma) \wedge t(p_{\Gamma'}) = t(p_{\Gamma \cap \Gamma'})$,
$\neg t(p_\Gamma) = t(p_{\overline{\Gamma}}), t(p_\Gamma) \leq t(p_{\Gamma'}) \Leftrightarrow t(p_{\overline{\Gamma} \cup \Gamma'}) = \mathbb{1}$,
$t(p_\Gamma) \rightarrow t(p_{\Gamma'}) = t(p_{\Gamma \rightarrow \Gamma'}), t(p_\Gamma) \leftrightarrow t(p_{\Gamma'}) = t(p_{\Gamma \leftrightarrow \Gamma'})$
(see Supplement 8.4.5).

The truth functions $t(p_j) = t(p_{\{j\}}) = t(\bigwedge_{i \in N_n} p_{ij_i}) = \bigwedge_{i \in N_n} t(p_{ij_i})$ with $t(p_j) \neq \emptyset$ are the atoms of $\mathcal{T}^*$. They form a partition of T:

$$t(p_j) \wedge t(p_k) = \emptyset \quad \text{for } j \neq k, \quad \bigwedge_{j \in K} t(p_j) = 1.$$

Finally it holds

$$\mathcal{T}^* = \{t(p) : p \in \mathcal{R}(\{p_{10}, ..., p_{nk_n}\}).$$

Proof

We omit the proof since it is similar to 8.4.4 and Supplement 8.4.5.

8.5.6 Properties of $\mathcal{T}^*$

Suppose $t(p_j) \neq \emptyset$ for all $j \in K$. Then again the list 7.2.5 of properties of $\mathcal{B}^*$ may be read as a list of properties of $\mathcal{T}^*$: Replace B_{ij} by $t(p_{ij})$, B_j by $t(p_j)$, B_Γ by $t(p_\Gamma)$.

8.5.7 Representation of Truth Functions out of $\mathcal{T}^*$ by Implicants and Implicates

In the preceding Sections 8.1, 8.2 and 8.4 we explicitely stated representations by implicants and implicates for sets, indicators and classes of propositions. In each case we used the general results of Chapter 7 concerning Boolean algebras. In the same way we obtain such representations for truth functions too. Therefore we omit the corresponding results and leave the translation from the general Boolean algebra $\mathcal{B}^*$ into the special truth function algebra $\mathcal{T}^*$ to the reader.

8.6 Some Related Models

In this section let us consider some models which may be obtained from the foregoing models by slight modifications.

8.6.1 Classes of Propositional Variables defined on Ω

Let Ω be a set. Then, for each $A \subseteq \Omega$, we may define a propositional variable $p_A(\omega)$ on Ω being true or false in such a way that

$$p_A(\omega) \text{ is } \begin{cases} \text{true} & \text{if } \omega \in A \\ \text{false} & \text{if } \omega \in \overline{A}. \end{cases}$$

Then $\{p_{A_1}(\omega), ..., p_{A_r}(\omega)\}$ is a partition of $\mathbb{T} = p_\Omega(\omega)$ if and only if $\{A_1, ..., A_r\}$ is a partition of Ω. As in 8.4 we may handle Boolean algebras of classes $(p_{A_\Gamma}(\omega))^*$ with $\Gamma \subseteq K$.

8.6.2 Truth Functions defined on Ω

To each $p_A(\omega)$ defined in 8.6.1 we define in an evident way a truth function $t(p_A(\omega))$ by

$$t(p_A(\omega)) = \begin{cases} T & \text{if } \omega \in A \\ F & \text{if } \omega \in \overline{A} \, . \end{cases}$$

Again $\{t(p_{A_1}(\omega)), ..., t(p_{A_r}(\omega))\}$ is a partition of $t(p_\Omega(\omega)) \equiv T$ if and only if $\{A_1, ..., A_r\}$ is a partition of Ω. As in 8.5 we may handle Boolean algebras of truth functions $t(p_{A_\Gamma}(\omega))$ with $\Gamma \subseteq K$.

8.6.3 Boolean functions of Restricted Variables

Usually Boolean functions are defined as functions $f : \{0,1\}^n \to \{0,1\}$, i.e. the Boolean n–tuples $(x_1, ..., x_n)$ takes all values out of $\{0,1\}^n = \bigtimes_{i=1}^{n} \{0,1\}$.

Now we ask for "Boolean" functions $f(x_{10}, ..., x_{1k_1}, ..., x_{n0}, ..., x_{nk_n})$ of $k_1 + 1 + ... + k_n + 1$ "Boolean" variables satisfying the restriction

$$\sum_{\kappa=0}^{k_1} x_{1\kappa} = \cdots = \sum_{\kappa=0}^{k_n} x_{n\kappa} = 1$$

(only one of the variables $x_{i0}, ..., x_{ik_i}$ takes 1, the other 0, $i \in N_n$).

The next theorem yields the set of all these functions.

8.6.3.1 Theorem

For $j = (j_1, ..., j_n) \in K = \bigtimes \{0, ..., k_i\}$ define the $(k_1 + 1 + \cdots + k_n + 1)$–tuple $x_{(j)}$ by

$$x_{(j)} := (y_{10}, ..., y_{1k_1}, ..., y_{n0}, ..., y_{nk_n}) \text{ with } y_{i\kappa} = \begin{cases} 1 & \text{for } i \in N_n, \kappa = j_i \\ 0 & \text{otherwise.} \end{cases}$$

Let $\Omega_K := \{x_{(j)} : j \in K\}$. Then, for the set $\{0,1\}^{\Omega_K}$ of all "Boolean" functions $f : \Omega_K \to \{0,1\}$, we have

$$\{0,1\}^{\Omega_K} = \{f_\Gamma : \Gamma \subseteq K\}$$

with

$$f_\Gamma(x) := f_\Gamma(x_{10}, ..., x_{nk_n}) := \sum_{j \in \Gamma} \prod_{i \in N_n} x_{ij_i},$$

and f_Γ is the indicator of the set $D_\Gamma := \{x_{(j)} \in \Omega_K : j \in \Gamma\} \subseteq \Omega_K, \Gamma \subset K$.
For $i \in N_n, \kappa \in \{0,...,k_i\}$, let $\Gamma_{i\kappa} := \{j \in K : j_i = \kappa\}$. Then $f_{\Gamma_{i\kappa}}(x) = x_{i\kappa}$.

Proof

By definition of $x_{(j)}$, for $j \in K$, we have

$$\prod_{i \in N_n} x_{ij_i} = \begin{cases} 1 & \text{for } x = x_{(j)} \\ 0 & \text{otherwise,} \end{cases}$$

i.e.

$$1_{\{x_{(j)}\}}(x) = \prod_{i \in N_n} x_{ij_i}.$$

By Definition of D_Γ, for $\Gamma \subseteq K$, we obtain $1_{D_\Gamma}(x) = \sum_{j \in \Gamma} \prod_{i \in N_n} x_{ij_i} =: f_\Gamma(x)$.
Clearly, for each $\Gamma \subseteq K, f_\Gamma \in \{0,1\}^{\Omega_K}$. Now suppose $g \in \{0,1\}^{\Omega_K}$. Let $\Gamma(g) := \{j \in K : g(x_{(j)}) = 1\} \subseteq K$. Obviously then $g(x) = 1_{D_{\Gamma(g)}}(x) = f_{\Gamma(g)}(x)$ and so $g \in \{f_\Gamma : \Gamma \subseteq K\}$.
Finally by Definition of $\Gamma_{i_0\kappa_0}$, for $i_0 \in N_n, \kappa_0 \in \{0,...,k_{i0}\}$, we obtain

$$f_{\Gamma_{i_0\kappa_0}}(x) = \sum_{\substack{j \in K \\ j_{i_0} = \kappa_0}} \prod_{i \in N_n} x_{ij_i} = x_{i_0\kappa_0} \sum_{\substack{j_i \in \{0,...,k_i\} \\ \text{for } i \neq i_0}} \prod_{\substack{i \in N_n \\ i \neq i_0}} x_{ij_i}$$

$$= x_{i_0\kappa_0} \prod_{\substack{i \in N_n \\ i \neq i_0}} \sum_{j_i=0}^{k_i} x_{ij_i} = x_{i_0\kappa_0} \prod_{\substack{i \in N_n \\ i \neq i_0}} 1 = x_{i_0\kappa_0}. \qquad \square$$

Remark

For $i \in N_n, \kappa \in \{0,...,k_i\}$ let

$$D_{i\kappa} := \{x_{(j)} \in \Omega_K : j_i = \kappa\} \subseteq \Omega_K.$$

Then from Theorem 8.1.1 and Theorem 8.2.3 we may deduce, that $\{f_\Gamma : \Gamma \subseteq K\}$ is the indicator algebra generated by the indicators $1_{D_{10}}, ..., 1_{D_{nk_n}}$, where

$$1_{D_{i\kappa}}(x) = f_{\Gamma_{i\kappa}}(x) = x_{i\kappa}.$$

We note that the list 7.2.5 of properties holds for $\{f_\Gamma : \Gamma \subseteq K\}$ – see 8.2.5.

8.6.4 Binary functions

Let $M = K$. Then the relation between the Boolean functions f_Γ and the binary functions 1_Γ (chapter 1) is the following: Define the variables $x_1, ..., x_n$ by

$$x_i := \sum_{j=0}^{k_i} x_{ij}, j \in N_n.$$

Then, for $i \in N_n$, holds

$$x_i = j \text{ if and only if } x_{ij} = 1, \; j \in \{0, ..., k_i\},$$

and we obtain

$$1_\Gamma(x_1, ..., x_n) = f_\Gamma(x_{10}, ..., x_{nk_n}).$$

8.7 Calculation of Elements of $\mathcal{B}^*$

Elements of a Boolean algebra

$$\mathcal{B}^* := \mathcal{B}(\{B_{10}, ..., B_{nk_n}\}) = \{B_\Gamma : \Gamma \subseteq K\}$$

according to Theorem 7.2.3 often are given as expressions obtained from some of the generating elements $B_{10}, ..., B_{nk_n}$ by any applications of the Boolean connectives $\vee, \wedge, \neg, \rightarrow, \leftrightarrow$ (with $B \rightarrow B' := \neg B \vee B', B \leftrightarrow B' := (B \wedge B') \vee (\neg B \wedge \neg B')$).

In some cases such expressions may be complicated or difficult to survey. Therefore we are interested in suitable methods to transform respectively to simplify such "Boolean expressions". We may ask e.g. for a $\Gamma \subseteq K$ so that the concerned expression equals B_Γ.

We will show that the Boolean functions treated in 8.6 are a good tool to transform and to simplify such Boolean expressions.

8.7.1 Definition

Let $A(B_{10}, ..., B_{nk_n})$ be an element of $\mathcal{B}^*$ given as an expression obtained from some of the elements $B_{10}, ..., B_{nk_n}$ by any applications of the connectives $\vee, \wedge, \neg, \rightarrow, \leftrightarrow$. Then we call $A(B_{10}, ..., B_{nk_n})$ a *Boolean expression out of* $\mathcal{B}^*$.

If we replace in $A(B_{10}, ..., B_{nk_n})$ each (occuring) $B_{\iota\kappa}$ by the corresponding Boolean variable $x_{\iota\kappa}$ according to 8.6.3 then we obtain a Boolean function given as

$A(x_{10},...,x_{nk_n})$, where the connectives $\vee, \wedge, \neg, \rightarrow, \leftrightarrow$ are defined by

$$\begin{aligned} y \vee y' &:= \max(y,y') = 1-(1-y)(1-y') = y+y'-y\cdot y', \\ y \wedge y' &:= \min(y,y') = y\cdot y', \\ \neg y &:= 1-y, \\ y \rightarrow y' &:= \neg y \vee y' = 1-y+y\cdot y', \\ y \leftrightarrow y' &:= (y\wedge y')\vee(\neg y \wedge \neg y') = 1-y-y'+2y\cdot y'. \end{aligned}$$

Moreover we have, for $i \in N_n$, $\sum_{k=0}^{k_i} x_{ik} = 1$ implying $x_{ik} \wedge x_{ik'} = 0$ for $k \neq k'$.

Example

Let $n=3, k_1=k_2=k_3=2$,

$$A(B_{10},...,B_{32}) = [(B_{10} \rightarrow B_{21}) \vee (B_{20} \wedge B_{31})] \wedge \neg(B_{21} \vee B_{22}).$$

The corresponding Boolean function is

$$A(x_{10},...,x_{32}) = [(x_{10} \rightarrow x_{21}) \vee (x_{20} \wedge x_{31})] \wedge \neg(x_{21} \vee x_{22}).$$

Using our "calculation rules" we obtain

$$A(x_{10},...,x_{32}) = x_{11}\cdot x_{20} + x_{12}\cdot x_{20} + x_{10}\cdot x_{20}\cdot x_{31} = x_{20}(x_{11}+x_{12}+x_{10}\cdot x_{31}),$$

and this may be written again as

$$\begin{aligned} A(x_{10},...,x_{32}) &= (x_{11}\wedge x_{20}) \vee (x_{12}\wedge x_{20}) \vee (x_{10}\wedge x_{20}\wedge x_{31}) \\ &= x_{20} \wedge (x_{11} \vee x_{12} \vee (x_{10}\wedge x_{31})). \end{aligned}$$

The following statement says that the eqality of two Boolean functions given as $A(x_{10},...,x_{nk_n})$ and $A'(x_{10},...,x_{nk_n})$ implies the equality of the corresponding Boolean expressions out of $\mathcal{B}^*$. We give it without proof.

8.7.2 Theorem

Let $A(B_{10},...,B_{nk_n})$ and $A'(B_{10},...,B_{nk_n})$ be Boolean expressions out of $\mathcal{B}^*$. Let $x_{10},...,x_{nk_n}$ be the Boolean variables according to 8.6.3. Then

$$A(x_{10},...,x_{nk_n}) = A'(x_{10},...,x_{nk_n})$$

implies

$$A(B_{10},...,B_{nk_n}) = A'(B_{10},...,B_{nk_n}).$$

Example

Let $A(B_{10},...,B_{32})$ given as before. Then with $A'(x_{10},...,x_{32}) = (x_{11} \wedge x_{20}) \vee (x_{12} \wedge x_{20}) \vee (x_{10} \wedge x_{20} \wedge x_{31}) = x_{20} \wedge (x_{11} \vee x_{12} \vee (x_{10} \wedge x_{31}))$ we obtain

$$\begin{aligned}&[(B_{10} \to B_{21}) \vee (B_{20} \wedge B_{31})] \wedge \neg(B_{21} \vee B_{22})\\ &\quad = (B_{11} \wedge B_{20}) \vee (B_{12} \wedge B_{20}) \vee (B_{10} \wedge B_{20} \wedge B_{31})\\ &\quad = B_{20} \wedge (B_{11} \vee B_{12} \vee (B_{10} \wedge B_{31})).\end{aligned}$$

Finally we give a decomposition theorem which is useful to transform a Boolean function $f(x_{10},...,x_{nk_n})$ given by any Boolean expression into a sum of products of the variables. Especially it can be used to determine the standard representation as $f_\Gamma(x_{10},...,x_{nk_n})$. We give it without proof too.

8.7.3 Theorem

Let $A(x_{10},...,x_{n0})$ be a Boolean Expression in the Boolean variables $x_{10},...,x_{nk_n}$ according to 8.6.3. Then, for each $L \subseteq N_n$, $L \neq \emptyset$

$$(8.7.4) \qquad A(x_{10},...,x_{nk_n}) = \sum_{\substack{j_i \in \{0,...,k_i\} \\ \text{for } i \in L}} \prod_{i \in L} x_{ij_i} A(x_{10},...,x_{nk_n})|_{x_{ij_i}=1 \text{ for } i \in L}.$$

wher $x_{ij_i} = 1$ means together $x_{i\kappa} = 0$ for $\kappa \neq j_i, \kappa \in \{0,...,k_i\}$.

8.7.5 Corollary

For each $i \in N_n$,

$$(8.7.6) \qquad A(x_{10},...,x_{nk_n}) = \sum_{j_i=0}^{k_i} x_{ij_i} A(x_{10},...,x_{nk_n})|_{x_{ij_i}=1}.$$

For the practical purpose it is recommendable to apply first (8.7.6) with $i = 1$. Obviously the expressions $A(x_{10},...,x_{nk_n})|_{x_{1j_1}=1}, j_1 \in \{0,...,k_1\}$, are expressions which depend only on the variables $x_{20},...,x_{nk_n}$. Thus we may repeate the procedure with the expressions

$$A_{1j_1}(x_{20},...,x_{nk_n}) := A(x_{10},...,x_{nk_n})|_{x_{1j_1}} = 1$$

and obtain

$$A_{1j_1}(x_{20},...,x_{nk_n}) = \sum_{j_2=0}^{k_2} x_{2j_2} A_{1j_1}(x_{20},...,x_{nk_n})|_{x_{2j_2}=1},$$

etc. We may end the procedure as soon as we obtain $A(x)$ as a sum of products each of them containing at most n factors. Moreover we may represent $A(x)$ as an $f_\Gamma(x), \Gamma \subseteq K$, if we continue the procedure until only products with exactly n factors are obtained. Now from Theorem 8.7.2 we obtain

$$A(B_{10}, ..., B_{nk_n}) = A_\Gamma.$$

We note that a more general decomposition theorem for elements of $\mathcal{B}^*$ is given by

$$A(B_{10}, ..., B_{nk_n}) = \bigvee_{\substack{j_i \in \{0,...,k_i\} \\ \text{for } i \in L}} \bigwedge_{i \in L} B_{ij_i} \wedge A(B_{10}, ..., B_{nk_n})|_{B_{ij_i}=\mathbb{1} \text{ for } i \in L}.$$

where $B_{ij_i} = \mathbb{1}$ means: Replace in $A(B_{10}, ..., B_{nk_n})$ B_{ij_i} by $\mathbb{1}$ (and so $B_{i\kappa}$ by $\mathbb{O}$ for $\kappa \neq j_i$) for $i \in L$.

Concluding Remark

In this Book we showed how binary functions can be represented in an optimal way by implicants and implicates, moreover that implicants and implicates of binary functions are useful to construct corresponding representations of discrete functions. Then we considered quite different systems as set algebras (used in probability theory), classes of propositions (belonging to propositional logic), further indicators, truth functions, and modified Boolean functions.

For all these systems we obtained representations by implicants or implicates in a simple way from the representations of the coordinated binary functions.

So we can simplify e.g. sets, propositions, indicators, truth functions, or Boolean functions submitted in form of complicated (Boolean) expressions. In Section 8.7 we gave some hints how to use Boolean functions to transform any Boolean elements as sets, classes of propositions etc.

Obviously for practical applications suitable computer programs are needed.

We did not take into consideration the complexity of such programs.

References

Böhme, G. (1981) Einstieg in die Mathematische Logik. Carl Hanser Verlag, München – Wien.

Davio, M., Deschamps, J.P. and Thayse, A. (1978) Discrete and Switching Functions. Mc Graw–Hill, New York and elsewhere.

Denis–Papin, M., Faure, R., Kaufmann, A. and Malgrange, Y. (1974) Theorie und Praxis der Booleschen Algebra. Vieweg, Braunschweig

Gaede, K.W. (1977) Zuverlässigkeit, Mathematische Modelle. Carl Hanser Verlag, München – Wien.

Hailperin, Th. (1976) Booles Logic and Probability. North–Holland Publishing Company, Amsterdam – New York – Oxford.

Höfle–Isphording, U. (1978) Zuverlässigkeitsrechnung, Einführung in ihre Methoden. Springer–Verlag, Berlin – Heidelberg – New York.

Störmer, H. (1983) Mathematische Theorie der Zuverlässigkeit, 2nd ed. R. Oldenbourg Verlag, München – Wien.

Wegener, I. (1987) The Complexity of Boolean Functions. John Wiley and Sons, New York and elsewhere; B.G. Teubner, Stuttgart.

Wilks, S.S. (1962) Mathematical Statistics. John Wiley and Sons, New York.

List of Symbols

Chapter 2

M_i, 3
M, 3
Γ, 3
$1_\Gamma(x)$, 3
$\overline{\Gamma}$, 3
P_i, 4
$C(P)$, 4
$C(P_1, ..., P_n)$, 4
$C(a)$, 5
N_n, 5
K, 5
$\overline{K}$, 5
$C(a^K, P^{\overline{K}})$, 5
a^K, 5
L, 6
P_i^*, 6
1_Γ, 7
$\mathcal{C}(\Gamma)$, 8
$\mathcal{C}_k(\Gamma)$, 8
$N_{i,\Gamma}(a^K, P^{\overline{K}})$, 10
$N_{i,\Gamma}^+(a^K, P^{\overline{K}})$, 12
$N_{i,\Gamma}'(a^K, P^{\overline{K}})$, 12
C, 12
$P_{(\rho)}$, 14
$\mathcal{C}_p(\Gamma)$, 17
$\mathcal{I}(\Gamma)$, 17
$\mathcal{I}_p(\Gamma)$, 17
$\mathcal{R}(\Gamma)$, 17
$\mathcal{R}_p(\Gamma)$, 17
$\mathcal{S}(A_1, ..., A_m)$, 18
$R(\mathcal{T})$, 18
$\mathcal{M}_\mu(\Gamma)$, 19
$\mathcal{M}_{p\mu}(\Gamma)$, 19
$D_\mu(m)$, 19
$R(\mathcal{S}(A_1, ..., A_m))$, 19

Chapter 3

$F(P)$, 23
$F(P_1, ..., P_n)$, 23
$F(\overline{a})$, 23
$F(\overline{a}^K, P^{\overline{K}})$, 23
$\mathcal{F}(\Gamma)$, 26
$\mathcal{F}_k(\Gamma)$, 26
$N_{i,\Gamma}^*(\overline{a}^K, P^{\overline{K}})$, 27
$N_{i,\Gamma}^{*+}(\overline{a}^K, P^{\overline{K}})$, 28
$N_{i,\Gamma}^{*'}(\overline{a}^K, P^{\overline{K}})$, 28
F, 29
$\mathcal{F}_p(\Gamma)$, 34
$\mathcal{I}^*(\Gamma)$, 34
$\mathcal{I}_p^*(\Gamma)$, 34
$\mathcal{R}^*(\Gamma)$, 34
$\mathcal{R}_p^*(\Gamma)$, 34
$\mathcal{M}_\mu^*(\Gamma)$, 35
$\mathcal{M}_{p\mu}^*(\Gamma)$, 35

Chapter 4

R_K, 36
L_ρ, 39
$R^{(0)}_{N_\mu}$, 39
$R^{(1)}_{N_\mu}$, 39
$R^{(1,1)}_{n_\mu}$, 40
$A^*_{\mu+1}$, 40
$A^{**}_{\mu+1}$, 40
$R^{(0)}_{N_\mu,\nu}$, 40
$R^{(1)}_{N_\mu,\nu}$, 40
$R^{(1,1)}_{N_\mu,\nu}$, 40
$A^*_{\mu+1,\nu}$, 40
$A^{**}_{\mu+1,\nu}$, 40

Chapter 5

f, 44
$1_{\{f=y_i\}}$, 44
$1_{\{f\geq y_i\}}$, 44
$1_{\{f\leq y_i\}}$, 44
a^*, 48
$G_{\geq y}$, 48
$G^*_{\geq y}$, 49
$G^*_{\leq y}$, 49
$G_{\leq y}$, 49
G_{y^*}, 51
G^*_y, 51
$I(y,c)$, 57
$\mathcal{C}_p(f)$, 58
$\mathcal{C}(f)$, 60
$\mathcal{C}_{yi}(f)$, 60
$\mathcal{C}_{y_ip}(f)$, 60
$\mathcal{I}(f)$, 62
$\mathcal{I}_p(f)$, 62
$\mathcal{R}(f)$, 62
$\mathcal{R}_p(f)$, 62
$\mathcal{M}_{i\kappa}(f)$, 63
$\mathcal{M}_{i\kappa p}(f)$, 63
$\mathcal{I}_i(f)$, 63
$\mathcal{I}_{ip}(f)$, 63
M_N, 66
M'_N, 67
$\mathcal{N}_{ij}(f)$, 67
$\mathcal{N}_{ijp}(f)$, 67
$I^*(y_i,F)$, 68
I^*, 69
$\mathcal{F}(f)$, 70
$\mathcal{F}_p(f)$, 70
$\mathcal{F}_{y_i}(f)$, 72
$\mathcal{F}_{y_ip}(f)$, 73
$\mathcal{I}^*(f)$, 76
$\mathcal{I}^*_p(f)$, 76
$\mathcal{R}^*(f)$, 76
$\mathcal{R}^*_p(f)$, 76
$\mathcal{M}^*_{i\kappa}$, 77
$\mathcal{M}^*_{i\kappa p}$, 77
M^*_N, 78
$M^{*'}_N$, 78
$\mathcal{N}^*_{ij}(f)$, 78
$\mathcal{N}^*_{ijp}(f)$, 78

Chapter 6

$S(x_1, ..., x_n)$, 80
Γ_κ, 81
1_{Γ_κ}, 82

Chapter 7

$\mathcal{B}$, 85
Ω, 86
$\mathbb{O}$, 86
$\mathbb{1}$,86
$\mathcal{A}$, 86
$\mathcal{C}$, 86
B_A, 86
$\mathcal{B}(\mathcal{C})$, 88
K, 89
Γ, 89
B_j, 89
B_Γ, 89
$\mathcal{B}(\{B_{10}, ..., B_{nk_n}\})$, 89
$\mathcal{B}^*$, 94
$C^*(K_1, ..., K_n)$, 95
C^*, 95
$F^*(K_1, ..., K_n)$, 99
F^*, 101
P, 102
p_j, 103
B_i^*, 104
p_{ik}, 105

Chapter 8

$\mathcal{A}^*$, 110
A_j, 110
A_Γ, 110
1_A, 118
$\mathcal{I}$, 118
$\mathcal{I}^*$, 119
$C_I(K_1, ..., K_n)$, 120
C_I, 120
$F_I(K_1, ..., K_n)$, 121
F_I, 122
$\mathcal{R}$, 123
p, q, 123
$\neg p$, 123
$p \vee q$, 123
$p \wedge q$, 123
$p \rightarrow q$, 123
$p \leftrightarrow q$, 123
$A(p_1, ..., p_r)$, 124
$B(p_1, ..., p_r)$, 124
F, 124
T, 124
$\sim$, 124
$\mathbb{F}$, 124
$\mathbb{T}$, 124
$\mathcal{R}(p_1, ..., p_r)$, 125
p^*, 126
$\Phi(\mathcal{R})$, 126
Φ^*, 128

p_Γ^*, 128
p_Γ, 128
$\mathcal{R}(\{p_{10}, ..., p_{nk_n}\})$, 129
$\Gamma(p)$, 130
$C_p(K_1, ..., K_n)$, 131
C_p, 131
$F_p(K_1, ..., K_n)$, 132
$\mathcal{T}(\mathcal{R})$, 133
$t(p)$, 133
$\mathcal{T}^*$, 135
$\mathcal{T}(\{t(p_{10}), ..., t(p_{nk_n})\}$, 135
$p_A(\omega)$, 136
$t(p_A(\omega))$, 137
$x_{(j)}$, 137
Ω_K, 137
$f_\Gamma(x)$, 137
$D_{i,\kappa}$, 138
$A(B_{10}, ..., B_{nk_n})$, 139
$A(x_{10}, ..., x_{nk_n})$, 140

Subject Index

A

algebras of classes of propositions, 126
anticube, 23
anticube element, 100
anticube indicator, 23
antitone discrete functions, 47
atoms, 89
attributes of objects, 81

B

Binary function, 1
Boolean algebra, 86
Boolean algebra generated by finite partitions, 89
Boolean expression, 139
Boolean function, 1
Boolean variable, 137

C

classes of objects, 81
classes of propositions, 126
classification of objects, 81
cube, 4
cube element, 95
cube indicator, 5

D

decomposition theorem, 141
discrete functions, 44

E

equivalent, 124
essential
    prime implicant, 16
    prime implicate, 32
event, 110
event algebra, 110

F

finite Boolean algebra. 84

I

implicant
    of a binary function, 7
    of a discrete function, 57
    of a Boolean element, 95
implicate
    of a binary function, 25
    of a discrete function, 68
    of a Boolean element, 101
independence, 104
indicator, 3, 118
indicator algebra, 118
infimum of Boolean elements, 85
interval, 47
isotone discrete function, 47

K

k–maximal, 13
k–minimal, 29

L

logical propositions, 123

M

maximal point, 48
maximplicate, 25

minimal cover, 19
minimal point, 48
minimal representation
by implicants, 16
by implicates, 32
minimplicant, 7
minterm, 5
monotone discrete function, 47

O

order relation, 85

P

parallel circuit, 5
partition, 88
partitions in propositional logic, 125
prime implicant, 13
prime implicate, 29
probability, 103
propositional calculus, 123
propositional logic, 123
propositions, 123

R

reduced representation
by implicants, 16
by implicates, 32
reduced set, 18
reduced system, 22
reduction methods, 36
reliability structure of technical systems, 80
representation of binary functions
by implicants, 15
by implicates, 30
by prime implicants, 15
by prime implicates, 30
representation of discrete functions
by implicants, 61
by implicates, 74
by prime implicants, 61
by prime implicates, 74
representation of Boolean elements
by implicants, 95
by implicates, 101
by prime implicants, 95
by prime implicates, 101
representation of classes of propositions
by implicants, 131
by implicates, 132
by prime implicants, 131
by prime implicates, 132
representation of indicators
by implicants, 121
by implicates, 122
by prime implicants, 121
by prime implicates, 122
representation of sets
by implicants, 111
by implicates, 114
by prime implicants, 111
by prime implicates,114
representation of truth functions
by implicants, 136
by implicates, 136
by prime implicants, 136
by prime implicates, 136

S

semimonotone discrete function, 52
series circuit, 5
set algebra, 86
subsystem, 81
supremum of Boolean elements, 85

system, 80
system function, 80

T

truth functions, 133
truth function algebra, 133
truth table, 118, 124

Vol. 236: G. Gandolfo, P.C. Padoan, A Disequilibrium Model of Real and Financial Accumulation in an Open Economy. VI, 172 pages. 1984.

Vol. 237: Misspecification Analysis. Proceedings, 1983. Edited by T.K. Dijkstra. V, 129 pages. 1984.

Vol. 238: W. Domschke, A. Drexl, Location and Layout Planning. IV, 134 pages. 1985.

Vol. 239: Microeconomic Models of Housing Markets. Edited by K. Stahl. VII, 197 pages. 1985.

Vol. 240: Contributions to Operations Research. Proceedings, 1984. Edited by K. Neumann and D. Pallaschke. V, 190 pages. 1985.

Vol. 241: U. Wittmann, Das Konzept rationaler Preiserwartungen. XI, 310 Seiten. 1985.

Vol. 242: Decision Making with Multiple Objectives. Proceedings, 1984. Edited by Y.Y. Haimes and V. Chankong. XI, 571 pages. 1985.

Vol. 243: Integer Programming and Related Areas. A Classified Bibliography 1981–1984. Edited by R. von Randow. XX, 386 pages. 1985.

Vol. 244: Advances in Equilibrium Theory. Proceedings, 1984. Edited by C.D. Aliprantis, O. Burkinshaw and N.J. Rothman. II, 235 pages. 1985.

Vol. 245: J.E.M. Wilhelm, Arbitrage Theory. VII, 114 pages. 1985.

Vol. 246: P.W. Otter, Dynamic Feature Space Modelling, Filtering and Self-Tuning Control of Stochastic Systems. XIV, 177 pages. 1985.

Vol. 247: Optimization and Discrete Choice in Urban Systems. Proceedings, 1983. Edited by B.G. Hutchinson, P. Nijkamp and M. Batty. VI, 371 pages. 1985.

Vol. 248: Plural Rationality and Interactive Decision Processes. Proceedings, 1984. Edited by M. Grauer, M. Thompson and A.P. Wierzbicki. VI, 354 pages. 1985.

Vol. 249: Spatial Price Equilibrium: Advances in Theory, Computation and Application. Proceedings, 1984. Edited by P. T. Harker. VII, 277 pages. 1985.

Vol. 250: M. Roubens, Ph. Vincke, Preference Modelling. VIII, 94 pages. 1985.

Vol. 251: Input-Output Modeling. Proceedings, 1984. Edited by A. Smyshlyaev. VI, 261 pages. 1985.

Vol. 252: A. Birolini, On the Use of Stochastic Processes in Modeling Reliability Problems. VI, 105 pages. 1985.

Vol. 253: C. Withagen, Economic Theory and International Trade in Natural Exhaustible Resources. VI, 172 pages. 1985.

Vol. 254: S. Müller, Arbitrage Pricing of Contingent Claims. VIII, 151 pages. 1985.

Vol. 255: Nondifferentiable Optimization: Motivations and Applications. Proceedings, 1984. Edited by V.F. Demyanov and D. Pallaschke. VI, 350 pages. 1985.

Vol. 256: Convexity and Duality in Optimization. Proceedings, 1984. Edited by J. Ponstein. V, 142 pages. 1985.

Vol. 257: Dynamics of Macrosystems. Proceedings, 1984. Edited by J.-P. Aubin, D. Saari and K. Sigmund. VI, 280 pages. 1985.

Vol. 258: H. Funke, Eine allgemeine Theorie der Polypol- und Oligopolpreisbildung. III, 237 pages. 1985.

Vol. 259: Infinite Programming. Proceedings, 1984. Edited by E.J. Anderson and A.B. Philpott. XIV, 244 pages. 1985.

Vol. 260: H.-J. Kruse, Degeneracy Graphs and the Neighbourhood Problem. VIII, 128 pages. 1986.

Vol. 261: Th.R. Gulledge, Jr., N.K. Womer, The Economics of Made-to-Order Production. VI, 134 pages. 1986.

Vol. 262: H.U. Buhl, A Neo-Classical Theory of Distribution and Wealth. V, 146 pages. 1986.

Vol. 263: M. Schäfer, Resource Extraction and Market Structure. XI, 154 pages. 1986.

Vol. 264: Models of Economic Dynamics. Proceedings, 1983. Edited by H.F. Sonnenschein. VII, 212 pages. 1986.

Vol. 265: Dynamic Games and Applications in Economics. Edited by T. Başar. IX, 288 pages. 1986.

Vol. 266: Multi-Stage Production Planning and Inventory Control. Edited by S. Axsäter, Ch. Schneeweiss and E. Silver. V, 264 pages. 1986.

Vol. 267: R. Bemelmans, The Capacity Aspect of Inventories. IX, 165 pages. 1986.

Vol. 268: V. Firchau, Information Evaluation in Capital Markets. VII, 103 pages. 1986.

Vol. 269: A. Borglin, H. Keiding, Optimality in Infinite Horizon Economies. VI, 180 pages. 1986.

Vol. 270: Technological Change, Employment and Spatial Dynamics. Proceedings 1985. Edited by P. Nijkamp. VII, 466 pages. 1986.

Vol. 271: C. Hildreth, The Cowles Commission in Chicago, 1939–1955. V, 176 pages. 1986.

Vol. 272: G. Clemenz, Credit Markets with Asymmetric Information. VIII, 212 pages. 1986.

Vol. 273: Large-Scale Modelling and Interactive Decision Analysis. Proceedings, 1985. Edited by G. Fandel, M. Grauer, A. Kurzhanski and A.P. Wierzbicki. VII, 363 pages. 1986.

Vol. 274: W.K. Klein Haneveld, Duality in Stochastic Linear and Dynamic Programming. VII, 295 pages. 1986.

Vol. 275: Competition, Instability, and Nonlinear Cycles. Proceedings, 1985. Edited by W. Semmler. XII, 340 pages. 1986.

Vol. 276: M.R. Baye, D.A. Black, Consumer Behavior, Cost of Living Measures, and the Income Tax. VII, 119 pages. 1986.

Vol. 277: Studies in Austrian Capital Theory, Investment and Time. Edited by M. Faber. VI, 317 pages. 1986.

Vol. 278: W.E. Diewert, The Measurement of the Economic Benefits of Infrastructure Services. V, 202 pages. 1986.

Vol. 279: H.-J. Büttler, G. Frei and B. Schips, Estimation of Disequilibrium Models. VI, 114 pages. 1986.

Vol. 280: H.T. Lau, Combinatorial Heuristic Algorithms with FORTRAN. VII, 126 pages. 1986.

Vol. 281: Ch.-L. Hwang, M.-J. Lin, Group Decision Making under Multiple Criteria. XI, 400 pages. 1987.

Vol. 282: K. Schittkowski, More Test Examples for Nonlinear Programming Codes. V, 261 pages. 1987.

Vol. 283: G. Gabisch, H.-W. Lorenz, Business Cycle Theory. VII, 229 pages. 1987.

Vol. 284: H. Lütkepohl, Forecasting Aggregated Vector ARMA Processes. X, 323 pages. 1987.

Vol. 285: Toward Interactive and Intelligent Decision Support Systems. Volume 1. Proceedings, 1986. Edited by Y. Sawaragi, K. Inoue and H. Nakayama. XII, 445 pages. 1987.

Vol. 286: Toward Interactive and Intelligent Decision Support Systems. Volume 2. Proceedings, 1986. Edited by Y. Sawaragi, K. Inoue and H. Nakayama. XII, 450 pages. 1987.

Vol. 287: Dynamical Systems. Proceedings, 1985. Edited by A.B. Kurzhanski and K. Sigmund. VI, 215 pages. 1987.

Vol. 288: G.D. Rudebusch, The Estimation of Macroeconomic Disequilibrium Models with Regime Classification Information. VII, 128 pages. 1987.

Vol. 289: B.R. Meijboom, Planning in Decentralized Firms. X, 168 pages. 1987.

Vol. 290: D.A. Carlson, A. Haurie, Infinite Horizon Optimal Control. XI, 254 pages. 1987.

Vol. 291: N. Takahashi, Design of Adaptive Organizations. VI, 140 pages. 1987.

Vol. 292: I. Tchijov, L. Tomaszewicz (Eds.), Input-Output Modeling. Proceedings, 1985. VI, 195 pages. 1987.

Vol. 293: D. Batten, J. Casti, B. Johansson (Eds.), Economic Evolution and Structural Adjustment. Proceedings, 1985. VI, 382 pages. 1987.

Vol. 294: J. Jahn, W. Krabs (Eds.), Recent Advances and Historical Development of Vector Optimization. VII, 405 pages. 1987.

Vol. 295: H. Meister, The Purification Problem for Constrained Games with Incomplete Information. X, 127 pages. 1987.

Vol. 296: A. Börsch-Supan, Econometric Analysis of Discrete Choice. VIII, 211 pages. 1987.

Vol. 297: V. Fedorov, H. Läuter (Eds.), Model-Oriented Data Analysis. Proceedings, 1987. VI, 239 pages. 1988.

Vol. 298: S.H. Chew, Q. Zheng, Integral Global Optimization. VII, 179 pages. 1988.

Vol. 299: K. Marti, Descent Directions and Efficient Solutions in Discretely Distributed Stochastic Programs. XIV, 178 pages. 1988.

Vol. 300: U. Derigs, Programming in Networks and Graphs. XI, 315 pages. 1988.

Vol. 301: J. Kacprzyk, M. Roubens (Eds.), Non-Conventional Preference Relations in Decision Making. VII, 155 pages. 1988.

Vol. 302: H.A. Eiselt, G. Pederzoli (Eds.), Advances in Optimization and Control. Proceedings, 1986. VIII, 372 pages. 1988.

Vol. 303: F.X. Diebold, Empirical Modeling of Exchange Rate Dynamics. VII, 143 pages. 1988.

Vol. 304: A. Kurzhanski, K. Neumann, D. Pallaschke (Eds.), Optimization, Parallel Processing and Applications. Proceedings, 1987. VI, 292 pages. 1988.

Vol. 305: G.-J.C.Th. van Schijndel, Dynamic Firm and Investor Behaviour under Progressive Personal Taxation. X, 215 pages. 1988.

Vol. 306: Ch. Klein, A Static Microeconomic Model of Pure Competition. VIII, 139 pages. 1988.

Vol. 307: T.K. Dijkstra (Ed.), On Model Uncertainty and its Statistical Implications. VII, 138 pages. 1988.

Vol. 308: J.R. Daduna, A. Wren (Eds.), Computer-Aided Transit Scheduling. VIII, 339 pages. 1988.

Vol. 309: G. Ricci, K. Velupillai (Eds.), Growth Cycles and Multisectoral Economics: the Goodwin Tradition. III, 126 pages. 1988.

Vol. 310: J. Kacprzyk, M. Fedrizzi (Eds.), Combining Fuzzy Imprecision with Probabilistic Uncertainty in Decision Making. IX, 399 pages. 1988.

Vol. 311: R. Färe, Fundamentals of Production Theory. IX, 163 pages. 1988.

Vol. 312: J. Krishnakumar, Estimation of Simultaneous Equation Models with Error Components Structure. X, 357 pages. 1988.

Vol. 313: W. Jammernegg, Sequential Binary Investment Decisions. VI, 156 pages. 1988.

Vol. 314: R. Tietz, W. Albers, R. Selten (Eds.), Bounded Rational Behavior in Experimental Games and Markets. VI, 368 pages. 1988.

Vol. 315: I. Orishimo, G.J.D. Hewings, P. Nijkamp (Eds.), Information Technology: Social and Spatial Perspectives. Proceedings, 1986. VI, 268 pages. 1988.

Vol. 316: R.L. Basmann, D.J. Slottje, K. Hayes, J.D. Johnson, D.J. Molina, The Generalized Fechner-Thurstone Direct Utility Function and Some of its Uses. VIII, 159 pages. 1988.

Vol. 317: L. Bianco, A. La Bella (Eds.), Freight Transport Planning and Logistics. Proceedings, 1987. X, 568 pages. 1988.

Vol. 318: T. Doup, Simplicial Algorithms on the Simplotope. VIII, 262 pages. 1988.

Vol. 319: D.T. Luc, Theory of Vector Optimization. VIII, 173 pages. 1989.

Vol. 320: D. van der Wijst, Financial Structure in Small Business. VII, 181 pages. 1989.

Vol. 321: M. Di Matteo, R.M. Goodwin, A. Vercelli (Eds.), Technological and Social Factors in Long Term Fluctuations. Proceedings. IX, 442 pages. 1989.

Vol. 322: T. Kollintzas (Ed.), The Rational Expectations Equilibrium Inventory Model. XI, 269 pages. 1989.

Vol. 323: M.B.M. de Koster, Capacity Oriented Analysis and Design of Production Systems. XII, 245 pages. 1989.

Vol. 324: I.M. Bomze, B.M. Pötscher, Game Theoretical Foundations of Evolutionary Stability. VI, 145 pages. 1989.

Vol. 325: P. Ferri, E. Greenberg, The Labor Market and Business Cycle Theories. X, 183 pages. 1989.

Vol. 326: Ch. Sauer, Alternative Theories of Output, Unemployment, and Inflation in Germany: 1960–1985. XIII, 206 pages. 1989.

Vol. 327: M. Tawada, Production Structure and International Trade. V, 132 pages. 1989.

Vol. 328: W. Güth, B. Kalkofen, Unique Solutions for Strategic Games. VII, 200 pages. 1989.

Vol. 329: G. Tillmann, Equity, Incentives, and Taxation. VI, 132 pages. 1989.

Vol. 330: P.M. Kort, Optimal Dynamic Investment Policies of a Value Maximizing Firm. VII, 185 pages. 1989.

Vol. 331: A. Lewandowski, A.P. Wierzbicki (Eds.), Aspiration Based Decision Support Systems. X, 400 pages. 1989.

Vol. 332: T.R. Gulledge, Jr., L.A. Litteral (Eds.), Cost Analysis Applications of Economics and Operations Research. Proceedings. VII, 422 pages. 1989.

Vol. 333: N. Dellaert, Production to Order. VII, 158 pages. 1989.

Vol. 334: H.-W. Lorenz, Nonlinear Dynamical Economics and Chaotic Motion. XI, 248 pages. 1989.

Vol. 335: A. G. Lockett, G. Islei (Eds.), Improving Decision Making in Organisations. Proceedings. IX, 606 pages. 1989.

Vol. 336: T. Puu, Nonlinear Economic Dynamics. VII, 119 pages. 1989.

Vol. 337: A. Lewandowski, I. Stanchev (Eds.), Methodology and Software for Interactive Decision Support. VIII, 309 pages. 1989.

Vol. 338: J.K. Ho, R.P. Sundarraj, DECOMP: an Implementation of Dantzig-Wolfe Decomposition for Linear Programming. VI, 206 pages. 1989.

Vol. 339: J. Terceiro Lomba, Estimation of Dynamic Econometric Models with Errors in Variables. VIII, 116 pages. 1990.

Vol. 340: T. Vasko, R. Ayres, L. Fontvieille (Eds.), Life Cycles and Long Waves. XIV, 293 pages. 1990.

Vol. 341: G.R. Uhlich, Descriptive Theories of Bargaining. IX, 165 pages. 1990.

Vol. 342: K. Okuguchi, F. Szidarovszky, The Theory of Oligopoly with Multi-Product Firms. V, 167 pages. 1990.

Vol. 343: C. Chiarella, The Elements of a Nonlinear Theory of Economic Dynamics. IX, 149 pages. 1990.

Vol. 344: K. Neumann, Stochastic Project Networks. XI, 237 pages. 1990.

Vol. 345: A. Cambini, E. Castagnoli, L. Martein, P. Mazzoleni, S. Schaible (Eds.), Generalized Convexity and Fractional Programming with Economic Applications. Proceedings, 1988. VII, 361 pages. 1990.

Vol. 346: R. von Randow (Ed.), Integer Programming and Related Areas. A Classified Bibliography 1984–1987. XIII, 514 pages. 1990.

Vol. 347: D.R. Insua, Sensitivity Analysis in Multi-objective Decision Making. XI, 193 pages. 1990.

Vol. 348: H. Störmer, Binary Functions and their Applications. VIII, 151 pages. 1990.